salt and pepper

jody vassallo

salt and pepper

photographs by deirdre rooney

w h i t e c a p

introduction

Since the beginning of time salt and pepper have been prized for their culinary properties. For years people have been effectively using salt as a preservative for a vast array of foods, have also been trading with it and in ancient times salt was even used as a form of currency.

Today cooks around the world use both salt and salted foods to enhance their recipes. In Asia pungent fish sauce, soy sauce and miso are favoured, in Africa and throughout the Middle East they prefer tangy preserved lemons and plump juicy olives, while Mediterranean kitchens would be lost without anchovies, capers, feta, halloumi and cured meats. Within this book you will find recipes that utilize such salted ingredients, which can all easily be purchased from a supermarket or delicatessen. I have also provided you with recipes of how to make some of the simpler ones at home.

Just as highly regarded in the culinary world is salt's table partner, pepper. Often referred to as the "King of Spices," it has long been used as a sacred offering to gods and also as a currency for paying rent and taxes. Like salt, pepper has had a long history in the kitchen, but its original job was disguising the flavour of ageing meats, rather than working as a preservative. The pepper chapter showcases the different types of pepper and highlights the diversity of flavours this spice can create. It introduces you to the idea of combining pepper with not just savoury foods but also using it as an ingredient within some truly delightful desserts. Not many people realize that pepper helps to bring out the sweetness in food, which makes it a wonderful spice to use with fruits such as rhubarb, apple and berries. Pepper is also renowned for its digestive properties so I have often used it in rich cream dishes.

Many people nowadays purchase good-quality salts and the freshest pepper but are never really sure as to exactly which salt should be used for cooking and what one can be sprinkled straight onto food. As for pepper, it can be a bit of a mystery – there are white and black peppers but what else is available and which one should be used when? This book will answer all the questions you have about how to use the most widely available salts and peppers. I guarantee it will have you lining up several varieties in your kitchen and cooking recipes just to use a certain type of salt or to demonstrate to your friends how versatile a particular colour of pepper can be. You will come to appreciate, as I have done, why these two table partners have been treasured since time began. I wonder which will be your favourite…

salt

salt varieties

SEA SALT FLAKES Large soft salt crystals made by boiling and evaporating salt water, this salt is suited to most types of cooking as it crumbles and dissolves easily. Mild in flavour and rich in minerals it can be added straight to vegetables and salads or used in rubs for meat, chicken and seafood.

CELERY SALT A fine salt that is combined with ground celery seed. Best known for its use in the drink Bloody Mary, it is also nice used in rubs for fish or sprinkled on sandwich fillings. It can be successfully added to sour cream or cream cheese to create a flavouring for dips too.

FLEUR DE SEL This is harvested in France from a single day's evaporation of salt crust on top of a salt pond. It is the least salty, purest part of the saline. Prized for its delicate flavour it is a great all-round salt. Especially good with white meats, I also like to add it to soups rather than using commercial stocks.

ROCK (COARSE SEA) SALT Firm salt granules that are best added to salt grinders. This salt is also good for rubbing into pork for crackling, pickling and baked fish. It is not suited to adding straight to food as the granules are too hard – they are better used in dishes with long cooking times, as this gives the crystals time to dissolve.

GREY SEA SALT (sel gris de Guerande) This mineral-rich salt from the Atlantic is still hand harvested. The harvester sweeps the top of the evaporating sea water to collect it. It has a subtle flavour and a grey hue. It is slightly wetter than other salts and tends to clump so I use it in dishes that will be cooked.

HIMALAYAN PINK SALT Fine pink salt sourced from dried-up inland seas from 200 million years ago, the pink colour comes from the iron that is trapped within the ground. It has a delicate flavour and is a good all-purpose salt. It looks pretty on the rim of margaritas or sprinkled on vegetables.

PINK SEA SALT FLAKES These pretty pink salt flakes are sourced from inland saline waters. The flakes have a subtle flavour, which makes this a good salt for the table and cooking. Because it flakes easily and looks pretty I use it where it can be seen – on baked potatoes or sprinkled on fried foods such as tempura.

FINE TABLE (COOKING) SALT A fine salt that has often had iodine added to it. Use sparingly as the flavour is concentrated. Do not substitute it for fleur de sel, grey sea salt or pink sea salt as they are for flavouring and this is for salting. This is good for pickles, brines and recipes needing large quantities of salt.

mediterranean rub

Put ⅓ cup fleur de sel (French sea salt), 2½ teaspoons dried Italian herbs and 1 teaspoon of finely grated lemon zest into a bowl and mix to combine. Use this as a rub mixed with a little olive oil for chicken, seafood and lamb. It can be also be used as a sprinkle for potatoes and barbecued meats.
Makes ⅓ cup

szechuan sprinkle

Roast 3 tablespoons of Szechuan peppercorns in a dry frying pan over a medium heat for 3 minutes or until fragrant. Remove and allow them to cool slightly. Transfer to a mortar and pestle and grind to a powder. Remove and stir in 3 tablespoons of pink sea salt flakes. This is great used as a sprinkle for an Asian-inspired meal – use it as you would salt and pepper and add it to stir fries, marinades and salads.
Makes ½ cup

indian spice rub

Put 1 tablespoon of grey sea salt (sel gris de Guerande), 1 teaspoon ground turmeric, 1 teaspoon ground coriander, 1 teaspoon cumin seeds, 1 teaspoon black mustard seeds, 1 teaspoon ground garam masala and ½ teaspoon dried chilli powder into a bowl and mix to combine. Use this rub sparingly on meat, chicken or fish. Alternatively, mix it with 3 tablespoons of yogurt or oil to create simple marinades for meat or poultry.
Makes ⅓ cup

asian salt sprinkle

Put 6 very finely shredded kaffir lime leaves, 1 teaspoon of dried chilli flakes and 1 tablespoon dried lemongrass into a small food processor or spice grinder and pulse quickly until finely chopped. Do not overprocess. Remove the kaffir lime mixture, stir through ⅓ cup sea salt flakes and 1 teaspoon grated palm sugar or brown sugar. Use this as a sprinkle for cooked prawns and seafood. You can also try adding some lime juice and groundnut (peanut) oil to the mixture to make a delicious marinade.
Makes ½ cup

gomashio – japanese sesame salt sprinkle

Put 1 tablespoon sea salt flakes and 7 tablespoons toasted sesame seeds into a surabachi (ridged Japanese mortar and pestle) if possible, or an ordinary mortar and pestle. Grind until roughly crushed. Do not overwork or the mixture will become a paste. This is delicious sprinkled on rice, soups and vegetables.
Makes ⅔ cup

preserved lemons

Used in the Middle East and North Africa preserved lemons are wonderful in both sweet and savoury recipes. It is just the peel that is used. Remove the pith and flesh and discard, then rinse the peel to remove the excess salt before using.

Preparation time: 20 minutes + 1 month standing, makes an 8-cup (2-litre) jar

2 lb (1 kg) thin-skinned lemons
⅔ cup sea salt flakes
2 bay leaves
½ teaspoon black peppercorns
juice of 2 lb (1 kg) lemons

Wash and scrub the lemons. Cut the lemons as if you were quartering them, taking care not to cut all the way through, and stuff the cavity of each lemon with salt. Fit the lemons, bay leaves and peppercorns into a clean 8-cup (2-litre) glass jar, pressing them down firmly. Leave unopened for 3–4 days.

Press the lemons down again and pour in the lemon juice, making sure the juice covers the lemons. Seal the jar and leave in a cool, dark place for 1 month. (The longer they are left the better the flavour will be.)

pan fried halloumi with lemon

Halloumi is a wonderful salty cheese with a stretchy, bouncy texture. Sheep's milk cheese, it is cooked in its own whey then seasoned with dried mint or cumin seeds and salt before being matured in a brine solution for up to 6 weeks.

Preparation time: 10 minutes, cooking time: 5 minutes, serves 4 as a snack

8 oz (250 g) halloumi
1 tablespoon olive oil
1 tablespoon lemon juice
½ tablespoon fresh flat-leaf parsley, roughly chopped
cracked black pepper
4 thick slices of Italian bread, to serve
1 tomato, cut into thick slices, to serve
1¾ oz (50 g) rocket (arugula), to serve

Cut the halloumi into thick slices. Heat the oil in a large frying pan, add the halloumi and cook over a medium heat for 3–5 minutes or until golden brown on both sides.

Add the lemon juice and parsley to the pan and season with cracked black pepper. Serve with bread, tomato and rocket (arugula), or add the cheese to your favourite salad.

gravlax

Originating in Sweden, gravlax is traditionally served to celebrate the arrival of spring. Some recipes add 1 tablespoon of vodka, so feel free if you wish to. You can make half the amount of gravlax by using 1 side of a salmon.

Preparation time: 10 minutes + 2 days curing, serves 12

2 x 1½ lb (750 g) salmon fillets with skin on
1 bunch dill (around ½ oz/15 g), roughly chopped
3½ oz (100 g) rock salt
¼ cup granulated sugar
2 tablespoons white peppercorns, crushed

Remove any bones from the salmon, using tweezers, then put the salmon skin-side down onto a large piece of plastic wrap.

Combine the dill, salt, sugar and peppercorns and spread the mixture over the flesh. Place the other fillet on top, skin-side facing up. Wrap the fish tightly in a couple of layers of plastic wrap and place onto a large baking tray. Weight the fish down with a chopping board and a foil-covered brick or several heavy cans. Refrigerate the fish for 2 days, turning them every 12 hours and draining off any excess liquid as you go.

Remove the salmon from the brining mixture, cut into paper thin slices and serve as you would smoked salmon – with bread, cream cheese and capers. It is also lovely with scrambled eggs, as part of a lunch with salad leaves or served simply with crackers.

* Gravlax will keep for up to 1 week in an airtight container in the refrigerator. If you do store it, keep it as a fillet and only slice it when you need to.

marinated feta

Traditionally made with sheep's milk, feta is now often produced using cow's milk. I like to use Bulgarian ewe's milk feta when I can find it as it is often creamier. Feta is matured in brine, which is where it gets its salty flavour from.

Preparation time: 5 minutes + 7 days marinating, makes 16 fl oz (500 ml)

2 cups feta cheese, cut into cubes
3 sprigs lemon thyme
5 juniper berries
1 tablespoon dried mixed (rainbow) peppercorns
½ cup (4 fl oz/125 ml) extra virgin olive oil
bruschetta, to serve

Put the feta into a sterilised jar, add the thyme, juniper berries and mixed peppercorns then top with the olive oil. Seal and set aside to marinate for 7 days.

* Refrigerate the marinated feta after opening it – the oil will harden it in the refrigerator so take it out a few hours before serving and allow it come up to room temperature. Marinated feta is delicious tossed through pasta, in salads, on top of pizzas, as part of a meze platter or simply accompanied by bruschetta as is shown here.

preserved and marinated olives

This recipe is for die-hard olive lovers who have access to fresh olives and also have time on their hands (I consider this process to be a labour of love). If you'd like to cheat a little you can buy preserved olives and then marinate them.

Preparation time: 15 minutes + 4 weeks soaking + overnight drying + 1 week preserving in brine + 1 week marinating, makes 1 lb (500 g)

1 lb (500 g) fresh olives
1 cup rock salt
2 bay leaves
zest of 1 lemon
1 teaspoon dried mixed (rainbow) peppercorns

MARINADE SUGGESTIONS:
Lemon, garlic and rosemary
2 cloves garlic, thinly sliced
zest and juice of 1 lemon
2 tablespoons rosemary sprigs

Chilli, lime and anchovy
1 teaspoon dried chilli flakes
zest of 2 limes
4 anchovies, chopped

To cure and preserve the olives, use a small knife to cut a slit in both sides of each olive (this helps to release the bitterness). Soak the olives in a large bowl of cold water for 4 weeks, changing the water every day. Taste after 2 weeks for green and 4 weeks for black – if they are still bitter continue the process for another 2 weeks. Once the olives are no longer bitter, spread them out on absorbent kitchen paper and allow to dry overnight.

Put the olives into a pan, add the salt, 10 cups (2.5 litres) of water, bay leaves, lemon zest, and peppercorns, bring to a boil, reduce the heat and simmer for 5 minutes. Then transfer the olives and brine into sterilised jars, seal and leave for 7 days before opening.

To marinate them, drain the preserved olives, cover with cold water and allow to stand for 1 hour. Drain well. Add whatever marinade ingredients you've chosen and mix to combine. Pack into sterilised jars, cover with olive oil, seal and allow to marinate for 7 days.

* Marinated olives will keep for up to 6 months in the refrigerator.

opposite: marinated feta, *following page:* preserved and marinated olives

salted spice and hazelnut dip

This dip is also known as dukkah. Use fresh spices, good bread and the best extra virgin olive oil you can afford when making it. Dukkah is also nice sprinkled over vegetables or meat dishes.

Preparation time: 5 minutes,
cooking time: 3 minutes, serves 4

4 tablespoons sesame seeds
2 tablespoons coriander seeds
1½ tablespoons cumin seeds
¾ oz (25 g) roasted hazelnuts, skins removed and chopped
1 tablespoon fleur de sel (French sea salt)
½ teaspoon freshly ground black pepper
fresh crusty bread, cut into large bite-size pieces, to serve
¼ cup (2 fl oz/60 ml) extra virgin olive oil, to serve

Dry roast the spices in a frying pan until they are fragrant. Put the roasted seeds, nuts and salt and pepper into a food processor or mortar and pestle and grind until they have been roughly chopped and combined. Cut the bread into large pieces and serve with the olive oil and dukkah.

* Dukkah will last up to 1 month if stored in an airtight container in the refrigerator.

sweet and salty cucumbers

These are delicious on sandwiches with roast beef or chicken. Alternatively, they also make a nice addition to a cheese or cold meat platter.

Preparation time: 15 minutes + 30 minutes cooling time + 1 week pickling,
cooking time: 10 minutes, makes 6 cups (1.5 litres)

2 lb (1 kg) firm cucumbers
2 cups (16 fl oz/500 ml) white wine vinegar
1 cup granulated sugar
3 tablespoons sea salt flakes
1 tablespoon pickling spice
1 tablespoon fresh dill, chopped

Thinly slice the cucumbers and put in a non-metallic bowl. Combine the vinegar, sugar, salt, pickling spice and dill in a saucepan and stir over a low heat to dissolve the sugar. Allow to cool slightly. Pour over the cucumbers and set aside to cool. Store for 1 week in the refrigerator in sterilised jars before serving.

boiled tea eggs with 5 spice sesame salt

I like to serve these eggs as part of an Asian dinner party, they just look so beautiful.

Preparation time: 10 minutes,
cooking time: 1 hour 10 minutes, serves 4

4 eggs
3 tablespoons black tea leaves
1 cinnamon stick
2 star anise
½ teaspoon sea salt flakes
¼ cup (2 fl oz/60 ml) dark soy sauce
½ teaspoon 5 spice powder
2 tablespoons toasted sesame seeds
2 tablespoons pink sea salt flakes

Put the eggs into a pan, cover with cold water and bring to a boil. Simmer for 10 minutes, drain then transfer to a bowl of iced water to cool. Gently tap the eggs until their shells are covered with cracks, return to the pan, add the tea, cinnamon, star anise, salt, soy sauce and 2 cups (16 fl oz/500 ml) of water. Simmer, covered, for 1 hour. Combine the 5 spice powder, sesame seeds and pink salt in a bowl. Remove the eggs and peel. Serve them standing in the 5 spice salt.

opposite: salted spice and hazelnut dip, *following pages:* sweet and salty cucumbers *(left)* and boiled tea eggs with 5 spice sesame salt *(right)*

salt and pepper squid

This recipe is the reason I could never give up fried food. As with all deep frying the secret to success is in the temperature of the oil. To test if it is ready stand a bamboo chopstick upright in the oil – bubbles should form around the chopstick.

Preparation time: 20 minutes,
cooking time: 10 minutes, serves 4

2 lb (1 kg) baby squid, cleaned and cut into rings
3 tablespoons grey sea salt (sel gris de Guerande)
3 tablespoons white peppercorns
2 teaspoons caster (superfine) sugar
2 cups cornflour (cornstarch)
4 egg whites, lightly beaten
groundnut (peanut) oil, for deep frying

Cut the squid hoods into rings and the tentacles in half. Put the salt, peppercorns and sugar into a mortar and pestle or spice grinder and grind to a fine powder. Transfer the mixture to a bowl, add the cornflour (cornstarch) and mix to combine.

Coat the squid in the egg whites, then toss to coat with the seasoned flour. Deep fry the squid in batches in a wok over a medium heat for 2 minutes or until crisp and golden brown. Drain on absorbent kitchen paper. The squid is good served with lime wedges.

pink salt pretzels

The pretzel originated in Germany and started off as a snack to be enjoyed with beer. If you want large, breakfast-size pretzels divide the dough into 4 instead of 8. Pink sea salt is high in minerals, which gives this recipe a unique flavour.

Preparation time: 30 minutes + 1 hour 10 minutes standing, cooking time: 15 minutes, makes 8

1 teaspoon standard dried yeast
¼ teaspoon granulated sugar
5 fl oz (150 ml) warm milk
1½ cups (6 oz/185g) strong bread flour
½ teaspoon sea salt flakes
1 oz (30 g) butter, melted
1 egg yolk, lightly beaten
1 tablespoon pink sea salt flakes
1 tablespoon rock salt

Preheat the oven to 375°F (190°C/Gas Mark 5). Put the yeast, sugar and warm milk into a small bowl and stir. Allow to stand for 10 minutes or until the yeast is foamy. Put the flour and salt into a bowl, make a well in the center, add the yeast mixture and butter and stir until the dough comes together. Turn out onto a lightly floured surface and knead for 10 minutes or until smooth and elastic. Transfer the dough to a lightly oiled bowl, cover with plastic wrap and allow to stand in a warm, draught-free area for 1 hour or until the dough has doubled in size.

Knock back the dough and knead it again on a lightly floured surface for 3 minutes. Divide the dough into 8 pieces. Roll each piece out into a rope about ¾ in (2 cm) thick, shape into a circle then knot into a pretzel shape. Arrange onto 2 baking trays lined with baking paper. Glaze with the beaten egg yolk then combine the 2 salts and sprinkle them over the top. Sprinkle with a little water then bake for 10–15 minutes or until crisp and golden. Transfer to a wire rack to cool before serving.

saffron salt cod fritters

I have to confess an absolute weakness for salt cod (baccalao) or any dish containing it. The fish is packed in salt, which preserves and dries it and also gives it an exquisite salty fish flavour.

Preparation time: 30 minutes + 24 hours soaking + 30 minutes chilling, cooking time: 25 minutes, makes 18 fritters

1 lb (500 g) salt cod (baccalao)
1 onion, thinly sliced
2 cloves garlic, sliced
1 bay leaf
pinch saffron threads
1⅔ cups (13 fl oz/400 ml) milk
½ cup (4 fl oz/125 ml) white wine
6½ oz (200 g) floury potatoes such as King Edward, peeled and chopped
1 teaspoon finely grated lemon zest
2 tablespoons fresh flat-leaf parsley, chopped
1 egg yolk
cracked black pepper
plain (all-purpose) flour, for dusting
olive oil, for shallow frying
lemon wedges, to garnish

Put the salt cod (baccalao) into a shallow dish, cover with cold water and allow to stand for 24 hours, changing the water every 6 hours. Drain.

Put the onion, garlic, bay leaf, saffron, milk and wine into a frying pan, bring almost to a boil, add the cod and cook, covered, for 15 minutes or until the fish is tender and starts to flake away from the skin. Remove the pan from the heat and set aside to cool slightly.

Remove the fish from the liquid and use a fork to flake the flesh away from the skin and bones. Cook the potatoes in a saucepan of boiling water until soft then drain and mash them. Add the fish, lemon zest, parsley, egg yolk and pepper and mix to combine.

Shape the mixture into balls (use 1 tablespoon per ball), place on a tray or plate lined with baking paper and refrigerate for 30 minutes.

Dust the fritters with the flour and flatten slightly. Heat the oil in a frying pan then shallow fry the fritters over a medium-high heat in batches of 6 for 2 minutes on each side or until crisp and golden and cooked through. Serve garnished with lemon wedges.

vodka pomegranate crush with salted stirrers

Pomegranates have been used for years in cocktails; they form the base of grenadine syrup. I find them a little fiddly to eat so I prefer to cut them in half and juice them. Some are tangier than others so test the drink for sweetness before serving.

Preparation time: 10 minutes + 1 hour drying time, serves 4

4 thick bamboo skewers

1 egg white, lightly beaten
2 tablespoons pink sea salt flakes
½ cup (4 fl oz/125 ml) vodka
2 tablespoons lime juice
1 teaspoon grated fresh ginger
2 tablespoons honey
4 pomegranates, halved
1 cup (8 fl oz/250 ml) cranberry juice

To make the salted stirrers, brush the bottom 4 in (10 cm) of the bamboo skewers with egg white. Put the salt on a plate and roll the bottom of the skewers in it before setting them aside to dry.

Fill the glasses with ice. Divide the vodka between 4 glasses, add the lime juice, ginger and honey and mix to combine.

Juice the pomegranates, strain and discard the seeds. Add the pomegranate and cranberry juices to the glasses before serving with the salted stirrers.

* Note: if pomegranates are out of season you can use pomegranate juice instead.

opposite: saffron salt cod fritters, *following pages:* vodka pomegranate crush with salted stirrers

olive and rosemary bread

This is a great bread to serve with cheese and wine. Good delicatessens will stock anchovy-stuffed green olives – make sure you pat them on some absorbent kitchen paper before chopping them in order to remove any excess oil.

Preparation time: 20 minutes + 2 hours 45 minutes standing, cooking time: 35 minutes, serves 6

1¼ lb (600 g) white bread pre-mix
½ cup (2 oz/60 g) fine semolina
1 teaspoons grey sea salt (sel gris de Guerande)
3 tablespoons olive oil
1¾ oz (50 g) pitted Kalamata olives, roughly chopped
1¾ oz (50 g) anchovy-stuffed green olives, roughly chopped
½ tablespoon fresh rosemary, chopped

Combine the flours and salt in a bowl, then make a well in the center. Add the oil and enough water to bring the dough together (it will probably need around 13 fl oz/ 375–400ml). Turn the dough out onto a lightly floured surface and knead for 10 minutes or until smooth. Place the dough in a lightly oiled bowl, cover with plastic wrap and allow to stand in a warm, draught-free place for 2 hours or until doubled in volume.

Preheat the oven to 450°F (230°C/Gas Mark 8). Knock back the dough and knead the olives and rosemary into the dough. Shape it into an oval, place it onto a baking tray lined with baking paper then cover it with a piece of lightly oiled plastic wrap and leave in a warm, draught-free area for 45 minutes.

Remove the plastic wrap and cut a few slits in the top of the loaf. Bake it for 15 minutes then reduce the heat to 400°F (200°C/Gas Mark 6) and bake it for 20 minutes longer or until the loaf sounds hollow when tapped. Transfer it to a wire rack to cool before serving.

tapenade

Anchovy is the name given to small silvery fish in the Mediterranean. The ones normally available in supermarkets have been filleted, cured in salt then packed in oil. Tapenade can be used as a spread or serve it with crackers or grissini (breadsticks).

Preparation time: 10 minutes, makes: 1 cup

4 oz (125 g) pitted Kalamata olives
¾ oz (25 g) anchovies
½ tablespoon capers
2 cloves garlic
1 tablespoon basil leaves
2 teaspoons lemon juice
¼ cup (2 fl oz/60 ml) extra virgin olive oil

Put the olives, anchovies, capers, garlic, basil and lemon juice into a food processor and process to form a smooth paste. Transfer the mixture to a bowl and gently stir through the oil. It is now ready to serve.

* The tapenade will keep, if it is in an airtight container, in the refrigerator for up to 1 month.

miso soup

White miso is the lightest and least salty of all the misos. Both this and dashi granules are available in Asian food stores. Miso soup can be eaten for breakfast, lunch or dinner and as a nutritious snack.

Preparation time: 15 minutes, cooking time: 10 minutes, serves 4

2 tablespoons white miso
1 teaspoon dashi granules (Japanese soup stock)
2½ oz (75 g) fresh or 2¾ oz (80 g) dried shiitake mushrooms
1 carrot, julienned
2 spring onions (scallions), sliced, to serve
6½ oz (200 g) soft (silken) tofu, diced, to serve

Put the miso, dashi and 4 cups (1 litre) of water into a pan, stir until the miso dissolves then cook over a medium heat until the miso is hot. Do not boil. Add the mushrooms and carrot to the pan and cook until the vegetables are soft.

Serve the soup topped with the spring onions (scallions) and tofu.

ham, pecorino pepato and egg pan fried sandwich

This is a delicious brunch or light snack recipe. Make sure you use proper sliced leg ham or you could use crisp cooked bacon – it is this ingredient that adds the saltiness to the recipe.

Preparation time: 10 minutes, cooking time: 25 minutes, serves 4

1 tablespoon olive oil
4 eggs
8 thick slices white bread
8 oz (250 g) sliced leg ham
6½ oz (200 g) grated Pecorino pepato cheese
3½ oz (100 g) grated Gruyère cheese
cracked black pepper
2¾ oz (80 g) butter

Heat the oil in a frying pan then fry the eggs until cooked to your liking.

Lay 4 slices of bread onto a clean work surface. Top with the ham, Pecorino, Gruyère and eggs and season with pepper. Finish with another slice of bread. Butter the top slice of bread and place the sandwich, buttered side down, into a frying pan over a medium heat. Cook for 5 minutes or until the underside is crisp and golden.

Butter the slice of bread now on top and turn the sandwich over. Weight it down with a plate and cook until the bottom slice is golden brown. Keep that sandwich warm while you cook the remaining sandwiches.

chargrilled prawns with a salty thai dipping sauce

The Thais call their fish sauce nam pla; it is made from salted and fermented dried fish or shrimp. Sounds unappetising I know, but it forms the foundation of their cuisine. Used as a seasoning it has a wonderful, pungent salty flavour. I love it!

Preparation time: 20 minutes,
cooking time: 10 minutes, serves 4

1 lb (500 g) raw prawns (shrimps), peeled and deveined with the tails left intact

For the dipping sauce:
1 teaspoon sesame oil
2 tablespoons nam pla (Thai fish sauce)
2 tablespoons lime juice
1 large red chilli, seeded and finely chopped
3 tablespoons grated palm sugar or brown sugar

Cook the prawns (shrimps) on a lightly oiled griddle (chargrill plate) or barbecue until pink and tender.

To make the dipping sauce, whisk together the sesame oil, fish sauce, lime juice, chilli and palm sugar until the sugar dissolves. The dish goes well with crisp Asian greens and steamed jasmine rice.

salted fish dip

Traditionally known as brandade, this is a delicious appetizer that may be served with bread or crackers or used as a filling for sandwiches. I like to fold a little mayonnaise or crème fraîche through after it has cooled.

Preparation time: 10 minutes + 24 hours soaking, cooking time: 25 minutes, serves 4

14 oz (450 g) salt cod (baccalao)
¼ cup (2 fl oz/60 ml) extra virgin olive oil
½ cup (4 fl oz/125 ml) single (light) cream
3 cloves garlic, crushed
cracked black pepper

Put the salt cod (baccalao) into a shallow dish, cover with cold water and allow to stand for 24 hours, changing the water every 6 hours. Drain.

Place the salt cod in a frying pan, cover with water and bring to a boil, reduce the heat and simmer for 10–15 minutes. Allow the fish to cool slightly then remove the bones and skin and flake into small pieces using a fork.

Heat half of the olive oil in a pan, add the salt cod and cook, stirring constantly, for 5 minutes. Gradually add the cream, garlic and the remaining oil, stirring continually until the mixture resembles mashed potato then season with the pepper.

margarita

When it comes to cocktails this one wins hands down. For a self-confessed salt junkie this is heaven. Crust the glasses just before serving or the salt will be too moist. You could try using pink sea salt to create a pretty summery drink.

Preparation time: 5 minutes + 1 hour cooling, serves 4

1 cup caster (superfine) sugar
3 cups ice
½ cup (4 fl oz/125 ml) tequila
½ cup (4 fl oz/125 ml) Cointreau
1 cup (8 fl oz/250 ml) lime juice
1 lime, cut into wedges
¼ cup (1½ oz/45 g) sea salt flakes

First you need to make a sugar syrup – place the caster (superfine) sugar and ½ cup (125ml/4fl oz) of water into a saucepan then cook over a low heat, stirring constantly until the sugar dissolves. Bring to a boil and cook over a high heat for 5 minutes or until it has reduced slightly. Set the syrup aside to cool for about 1 hour before using.

Put the ice, tequila, Cointreau, sugar syrup and lime juice into a blender and blend until smooth. Rub the edges of 4 margarita glasses with the lime wedges. Spread the salt out onto a saucer, invert the glasses and press them lightly into the salt. Pour the margarita into them and serve straightaway.

opposite: salted fish dip, *following page:* margarita

olive, tomato and bacon corn bread

This bread is the perfect accompaniment to a big bowl of soup. Don't, whatever you do, use the cheap supermarket pitted olives. Serve the bread warm with lashings of butter – it is also lovely cold (if it lasts that long)!

Preparation time: 15 minutes, cooking time: 45 minutes, serves 6–8

1¼ cups (5 oz/255 g) self-raising flour
1 tablespoon caster (superfine) sugar
2 teaspoons baking powder
1 teaspoon fleur de sel (French sea salt)
¾ cup (3½ oz/110 g) polenta (cornmeal)
½ cup (2 oz/60 g) grated Cheddar cheese
1 teaspoon fresh thyme, chopped
3½ oz (100 g) semi-dried tomatoes, finely chopped
2½ oz (75 g) pitted Kalamata olives, roughly chopped
3½ oz (100 g) lean bacon, chopped
2 eggs, lightly beaten
1 cup (8 fl oz/250 ml) buttermilk
⅓ cup (2¾ fl oz/80 ml) olive oil
1 tomato, sliced
3 sprigs thyme

Preheat the oven to 350°F (180°C/Gas Mark 4). Grease and line an 8 in x 4 in (20 cm x 10 cm) loaf tin. Sift the flour, sugar, baking powder and salt into a bowl. Add the polenta (cornmeal), cheese, fresh thyme, tomatoes, olives and bacon, then mix to combine. Make a well in the center. Whisk together the eggs, buttermilk and oil and pour into the well in the dry ingredients. Mix to combine. Pour the mixture into the loaf tin.

Garnish the top with the tomato slices and thyme sprigs. Bake for 45 minutes or until a skewer comes out clean when inserted into the center.

* If stored wrapped in plastic in an airtight container the bread will keep for 3 days. It is also delicious toasted.

spaghetti with chorizo, anchovies, chilli and garlic

I like to serve this pasta for lunch as it is a simple, easy-to-prepare meal. If you find anchovies too salty soak them in cold water or milk for 20 minutes before using. If you buy the anchovies in oil you can use the oil in the recipe to add more flavour.

Preparation time: 10 minutes, cooking time: 20–25 minutes, serves 4

13 oz (400 g) spaghetti
2 tablespoons olive oil
1 chorizo sausage, cut into thin strips
3 cloves garlic, crushed
4 spring onions (scallions), thinly sliced
1 teaspoon dried chilli flakes
6 anchovies, roughly chopped
2 tablespoons fresh parsley, chopped
2 tablespoons lemon juice
1¾ oz (50 g) butter
cracked black pepper

Cook the pasta in a large saucepan of rapidly boiling water until *al dente*. Drain, reserving a little of the cooking liquid.

Heat the oil in a frying pan, add the chorizo and cook over a medium heat for 3 minutes or until crisp. Add the garlic and spring onions (scallions) and cook over a medium heat for 3 minutes or until the spring onions are soft. Add the chilli and anchovies and cook for a further 3 minutes or until heated through. Add the parsley, lemon juice and butter and cook until the butter melts. Toss through the pasta and season with cracked black pepper before serving.

salty vietnamese clay pot pork

Nuoc man is the name the Vietnamese give to their salty fish sauce. The combination of sweet and salty works so well in this recipe. Fish and chicken are also delicious cooked this way – simply adjust the cooking time accordingly.

Preparation time: 15 minutes,
cooking time: 45 minutes, serves 4

4 tablespoons grated palm sugar or brown sugar
4 tablespoons nuoc man (Vietnamese fish sauce)
2 tablespoons chopped shallots
1 teaspoon cracked black pepper
1 lb (500 g) pork fillet, sliced
4 hard-boiled eggs, halved
4 spring onions (scallions), sliced, to serve
steamed rice, to serve

Heat the sugar and fish sauce in a clay pot or heatproof casserole dish, stirring constantly over a low heat until the sugar dissolves. Add the shallots and pepper and cook for 5 minutes. Add the pork, cover and simmer for 20 minutes, stirring occasionally. Add the hard-boiled eggs and cook, uncovered, for 5 minutes or until the eggs are heated through.

Serve the dish sprinkled with spring onions (scallions) and accompanied with steamed rice.

chicken with sweet chilli salt crust

These are wonderful bite-size snacks to serve with drinks. The sugar, salt and chilli work well together – the salt and sugar help temper the chilli. You can replace the chicken with prawns or squid, and simply adjust the cooking time.

Preparation time: 10 minutes, cooking time: 10 minutes, serves 6–8 as a snack

1 lb (500 g) chicken breast fillets
3 tablespoons pink sea salt flakes
2 teaspoons chilli powder
2 teaspoons caster (superfine) sugar
⅔ cup cornflour (cornstarch)
2 egg whites, lightly beaten
groundnut (peanut) oil, for deep frying

Cut the chicken breast into bite-size pieces. Roughly crush the salt in a mortar and pestle, transfer to a bowl, add the chilli, sugar and cornflour (cornstarch) and mix well to combine. Dip the chicken into the egg whites in batches, then roll in the chilli salt mixture and shake off any excess.

Deep fry the chicken in the oil, in batches, for 3–5 minutes or until crisp, golden and tender. Drain on absorbent kitchen paper. It is best served hot, topped with coriander (cilantro) leaves and accompanied by your choice of dipping sauce. I like to use sweet chilli sauce.

chinese beef with salted black bean sauce

Salted black beans are also known as fermented black beans. Many books will tell you to rinse the beans before using but I don't find this to be necessary (unless you want to reduce the saltiness). If you cannot find shao hsing wine use a dry sherry.

Preparation time: 15 minutes, cooking time: 15 minutes, serves 4

2 tablespoons groundnut (peanut) oil
1 lb (500 g) rump steak, thinly sliced
8 spring onions (scallions), thinly sliced
1 tablespoon grated fresh ginger
1 red pepper (capsicum), thinly sliced
2 cloves garlic, chopped
2 tablespoons salted black beans
3 tablespoons shao hsing wine (Chinese rice wine)
2 teaspoons caster (superfine) sugar
2 tablespoons oyster sauce
⅓ cup (2¾ fl oz/80 ml) chicken stock
½ teaspoon sesame oil
1 tablespoon cornflour (cornstarch)

Heat the oil in a wok, add the steak and stir fry over a high heat for 5 minutes or until browned. Add the spring onions (scallions), ginger, red pepper (capsicum), garlic and black beans and stir fry for 2 minutes or until the pepper is soft. Next, add the wine, sugar, oyster sauce, stock and sesame oil and cook until heated through.

Blend the cornflour (cornstarch) with 1 tablespoon of water and stir into the sauce. Cook, stirring constantly, until the sauce boils and thickens. Serve immediately. This dish is nice simply accompanied by steamed rice.

 opposite: chicken with sweet chilli salt crust, *following page:* chinese beef with salted black bean sauce

roast beef fillet in salt crust

Meat cooked in this way is so juicy and succulent. It is really important not to wrap the meat in advance or the pastry will be soggy. Also ensure there are no holes in the pastry as you want the meat to steam inside and keep the juices in.

Preparation time: 40 minutes + 20 minutes cooling + 30 minutes resting,
cooking time: 30 minutes for medium rare,
serves 4–6

2 tablespoons extra virgin olive oil
1 oz (30 g) butter
2 lb (1 kg) beef fillet
½ teaspoon cracked black pepper
1 tablespoon fresh rosemary, chopped

For the salt crust:
1 lb (500 g) fine table salt
4 cups plain (all-purpose) flour
2 tablespoons fresh lemon thyme, chopped

Heat the oil and butter in a large frying pan, add the beef and cook over a medium heat for 5 minutes, turning a couple of times until browned all over. Sprinkle with pepper and rosemary and set aside to cool while you prepare the pastry.

Put the salt, flour and thyme into a bowl. Gradually add in 1½ cups (12 fl oz/375 ml) of cold water and mix to form a soft dough. Rest for 30 minutes wrapped in plastic wrap.

Preheat the oven to 425°F (220°C/Gas Mark 7). Roll out the dough on a lightly floured surface until it is ¼ in (5 mm) thick. Sit the cooled beef onto the pastry and bring the edges together before pinching to seal. Bake for 30 minutes for medium rare, 40 minutes for medium, 55 minutes for well done. Allow the meat to rest for 20 minutes before carving. The dish is best served with roasted vegetables and gravy.

spanokopita

There are many different recipes for this Greek favourite. I particularly like this one as it caramelizes the onion and gives the pie a lot of extra flavour. Although the 1kg of spinach looks a lot it is correct, as it cooks down to nothing.

Preparation time: 15 minutes, cooking time: 1 hour, serves 4–6

2 lb (1 kg) English spinach
sea salt flakes and cracked black pepper
½ cup (4 fl oz/125 ml) olive oil
3 onions, finely chopped
1⅔ cups feta cheese, mashed
2 tablespoons chopped fresh flat-leaf parsley
2 tablespoons chopped fresh dill
10 sheets filo pastry

Preheat the oven to 350°F (180°C/Gas Mark 4). Lightly grease a 12 in x 8 in (30 cm x 20 cm) ovenproof dish. Steam the spinach until it wilts, then season it with the salt and pepper.

Heat half the oil in a large frying pan, add the onions and cook over a medium heat for 10 minutes or until the onions are golden brown. Remove from the heat and fold through the spinach, feta and herbs.

Line the ovenproof dish with 1 sheet of filo, brush lightly with the remaining oil and top with another sheet – continue layering and brushing with oil until you have used half the pastry. Top this with the spinach filling. Finish by layering the remaining pastry on top, brushing each sheet with oil as before. Trim the edges with a sharp knife and score the top of the pastry. Bake for 45 minutes or until crisp and golden brown. Allow to stand for 5 minutes before cutting.

brining solution for meat, fish and chicken

This recipe can be doubled to make a delicious brine for Christmas turkey – the brine helps to keep the bird's flesh lovely and moist. Shoulder or leg of pork is also delicious cooked this way.

Preparation time: 15 minutes + 48 hours brining, cooking time: 1 hour 20 minutes, serves 4–6

1 cup rock salt
½ cup brown sugar
1 bay leaf
5 cloves garlic
3 all-spice berries
3 juniper berries
3 lb (1.5 kg) chicken
olive oil, for drizzling

Put 20 cups (5 litres) of cold water, the salt, sugar, bay leaf, garlic and berries into a non-metallic bowl or container and mix well to combine and dissolve the sugar. Add the chicken and weight it down with a plate to ensure it is fully submerged in the brining solution. Chill for 48 hours.

Preheat the oven to 400°F (200°C/Gas Mark 6). Put the chicken onto an oven rack, drizzle with a little oil and bake for 15 minutes or until golden. Reduce the oven temperature to 350°F (180°C/Gas Mark 4) and bake the chicken for a further 1 hour or until the juices run clear when a skewer is inserted into the thickest part of the meat. The chicken can be served with roasted vegetables or salad and mayonnaise.

pissaladiere

Caramelizing the onions is an essential part of this recipe as the sweetness of the onions balance the saltiness of the anchovies. I find the traditional pizza base a little heavy so I have opted to replace it with a flaky puff pastry, which also saves time!

Preparation time: 10 minutes, cooking time: 50 minutes, makes 1 pizza (approx 4 servings)

3 tablespoons olive oil
1½ lb (750 g) onions, thinly sliced
2 ripe tomatoes, chopped
sea salt and cracked black pepper
1 sheet puff pastry (12 in/30 cm square)
12 anchovies
12 pitted Kalamata olives

Preheat the oven to 400°F (200°C/Gas Mark 6). Heat the oil in a large frying pan, add the onions, cover and cook, stirring occasionally, over a medium heat for 15 minutes or until golden brown – do not allow the onions to burn. Add the tomatoes and salt and pepper and cook for 10 minutes or until all the liquid from the tomatoes has been evaporated.

Roll out the pastry on a lightly floured surface then spread the onion mixture over the top, leaving a ¾ in (2 cm) border. Arrange the anchovies in a criss-cross pattern over the onions and put an olive into the center of each section. Place onto a baking tray lined with baking paper and bake for 25 minutes or until the pastry is golden and risen and the base is crisp.

opposite: chicken cooked in brining solution, *following page:* pissaladiere

roast vegetable salad with preserved lemon vinaigrette

This salad is best served warm and makes a good alternative to the standard vegetables served as part of a roast dinner. You can try changing the cheese – Camembert also works well.

Preparation time: 20 minutes,
cooking time: 40 minutes, serves 4–6

10 oz (300 g) butternut squash, cut into large pieces
1 lb (500 g) baby potatoes
6½ oz (200 g) baby onions
8 oz (250 g) baby carrots
10 oz (300 g) small parsnips
1 tablespoon olive oil
grey sea salt (sel gris de Guerande)
cracked black pepper
1 clove garlic, crushed
1 teaspoon Dijon mustard
1 tablespoon finely shredded, rinsed preserved lemon (see page 14)
2 tablespoons lemon juice
¼ cup extra virgin olive oil
3½ oz (100 g) goat's cheese, crumbled, to serve
sprigs of fresh rosemary, to serve

Preheat the oven to 400°F (200°C/Gas Mark 6). Put the vegetables into a large bowl, add the oil, salt and pepper and toss to coat the vegetables in the oil. Transfer to a large baking dish. Bake for 40 minutes or until the vegetables are tender. Remove from the oven.

In the meantime, whisk together the garlic, mustard, preserved lemon, lemon juice and olive oil. Pour over the roasted vegetable salad and gently toss to combine.

Serve topped with the crumbled goat's cheese and sprigs of fresh rosemary.

fish baked in salt crust with caperberry sauce

Don't be put off by the look of this recipe, I can see you questioning whether 2kg (4lb) of salt is a misprint – it is correct. The fish steams inside the salt and the result is amazing (don't eat the skin, though, as it will be far too salty).

Preparation time: 10 minutes,
cooking time: 30–40 minutes, serves 4–6

4 lb (2 kg) whole fish (salmon or sea bass), cleaned but not scaled, or 4 small fish for individual portions
4 lb (2 kg) fine table salt
5 egg whites

For the sauce:
12 caperberries, halved
½ cup (4 fl oz/125 ml) white wine
3 spring onions (scallions), sliced
3½ oz (100 g) butter, chopped
zest 1 lemon

Preheat the oven to 400°F (200°C /Gas Mark 6). Rinse and pat the fish dry with absorbent kitchen paper. Mix the salt with the egg whites (until it looks like wet sand) then spread half the salt mixture over the bottom of a baking dish large enough to fit the fish. Arrange the fish on top. Cover it with the remaining salt, pressing down to ensure there are no holes.

Bake the fish for 30–40 minutes or until a skewer inserted into the center of the fish comes out hot. Towards the end of the cooking time, make the sauce. Put the caperberries and white wine into a frying pan, bring to a boil and cook for 5 minutes, then add the spring onions (scallions), butter and lemon zest and cook for 3 minutes or until the butter melts and the spring onions soften.

Remove the fish from the oven, crack the salt crust with a large spoon and lift it away. Transfer the fish to a large plate and remove the skin. Serve it with the caperberry sauce.

potato, fennel and anchovy bake

This Swedish dish is also known as Janssons temptation. I have adapted it slightly to include fennel, but you can leave it out if it is out of season. The sweetness of the fennel, the subtleness of the potato and the saltiness of the anchovies make this my idea of the perfect comfort food.

Preparation time: 10 minutes + 10 minutes optional soaking,
cooking time: 1 hour 10 minutes,
serves 4–6 as a side dish

15 anchovies, chopped
1¾ oz (50 g) butter
1 onion, thinly sliced
1 fennel bulb, thinly sliced
5 potatoes, julienned
2 cups (16 fl oz/500 ml) single (light) cream

Preheat the oven to 400°F (200°C/Gas Mark 6). Soak the anchovies in water or milk for 10 minutes if you would like to reduce a little of the saltiness.

Heat the butter in a frying pan, add the onion and fennel and cook over a medium heat for 10 minutes or until soft and golden.

Arrange half the potatoes in the base of a 12 in x 8 in (30 cm x 20 cm) ovenproof dish, top with the onion and fennel mixture, anchovies and the remaining potatoes. Pour over the cream and bake for 1 hour or until the potato is soft and the top is golden.

veal saltimbocca

Italians call their ham "prosciutto." This type of ham has been seasoned, salt cured and air dried, not smoked. Make sure you purchase it from a delicatessen that will slice it into paper-thin slices otherwise this dish will be tough and chewy.

Preparation time: 15 minutes, cooking time: 20 minutes, serves 4

4 thin veal steaks
8 thin slices prosciutto
8 sage leaves
2 oz (60 g) butter
cracked black pepper
½ cup (4 fl oz/125 ml) dry white wine
steamed tender stem broccoli or purple sprouting broccoli, to serve

Flatten the veal steaks between 2 sheets of plastic wrap using a meat mallet.

Place 2 pieces of prosciutto and 2 sage leaves on top of each flattened veal steak, and secure with cocktail sticks (toothpicks).

Heat the butter in a large frying pan, add the steaks in batches and cook until just tender, then season with cracked black pepper. Add the wine to the pan, and cook over a high heat for 5 minutes or until the sauce has reduced slightly. Serve with steamed tender stem broccoli or purple sprouting broccoli.

feta chicken with honey caper sauce

This recipe is the perfect combination of sweet and salty. The hint of honey in the sauce works to soften the sharpness of the feta and the capers.

Preparation time: 25 minutes, cooking time: 20 minutes, serves 4

4 chicken breast fillets
3½ oz (100 g) feta cheese, crumbled
1 tablespoon fresh oregano, chopped
cracked black pepper
2 teaspoons olive oil
1 teaspoon grated lemon zest

For the sauce:
1 tablespoon olive oil
1 oz (30 g) butter
2 cloves garlic, crushed
2 tablespoons salted capers, rinsed
1 cup (8 fl oz/250 ml) verjuice
1 teaspoon honey
1 tablespoon fresh parsley, chopped

Preheat the oven to 350°F (180°C/Gas Mark 4). Cut a slit in the side of each chicken breast to form a pocket, taking care not to cut through to the other side. Put the feta, oregano, pepper, oil and lemon zest into a bowl and mix to combine. Divide the filling between the chicken breasts and secure the pockets using cocktail sticks (toothpicks).

Heat the oil and butter in a frying pan and cook the chicken breasts for 3 minutes on each side or until golden brown. Transfer to a baking tray and bake for 10–15 minutes or until tender. Meanwhile, place the garlic and capers in the frying pan and cook for 2 minutes. Add the verjuice and honey and cook over a medium-high heat for 10 minutes or until the sauce has reduced by half. Remove from the heat and stir through the parsley.

Remove the chicken breasts from the oven, cut into thick slices and pour the sauce over the top. This dish is nice served with a green salad.

opposite: veal saltimbocca, *following page:* feta chicken with honey caper sauce

preserved lemon granita

This granita is lovely served as a palate cleanser or dessert after a fish or seafood meal. It sounds time-consuming scraping 6 times, but it is a simple process and necessary if you want nice small ice crystals. Stand it at room temperature for 5 minutes before scraping to make your job easier.

Preparation time: 15 minutes + 8 hours freezing, serves 4

2 cups (16 fl oz/500 ml) lemon juice
½ cup caster (superfine) sugar
2 tablespoons chopped, rinsed preserved lemon (see page 14)
zest of 1 lemon

Put the lemon juice, sugar, preserved lemon and lemon zest into a saucepan and bring the mixture to a boil. Remove from the heat and allow to cool. Pour the mixture into a shallow non-stick metal container, cover and freeze for 2 hours or until the edges are firm. Remove the container from the freezer and scrape the mixture with a fork to break up the crystals then return to the freezer for 1 hour. Repeat the scraping and freezing 5 times over the next 5 hours or until the ice crystals are even in size.

Scrape the granita with a fork to break it up just before serving.

passion fruit and preserved lemon brûlées

The most important thing to remember when making any baked custard is not to overcook it. It should still be wobbly in the center when removed from the oven. Overcooking will cause it to separate and become watery.

Preparation time: 15 minutes, cooking time: 25 minutes + 4 hours chilling, serves 4

1 cup caster (superfine) sugar
4 eggs
1 cup (8 fl oz/250 ml) single (light) cream
6½ fl oz (200 ml) passion fruit pulp
1 tablespoon finely shredded, rinsed preserved lemon (see page 14)
2 teaspoons grated lemon zest
¼ cup granulated sugar

Preheat the oven to 315°F (160°C/Gas Mark 2–3). Put the caster (superfine) sugar, eggs, cream, passion fruit pulp, preserved lemon and lemon zest into a bowl and whisk to combine. Pour the mixture into four 1-cup (8-fl oz/250-ml) ramekins set in a baking dish then pour in enough boiling water to come halfway up the sides of the ramekins.

Cook for 25 minutes or until the custard is just set. Remove from the dish and allow to cool. Cover with plastic wrap and chill for 4 hours before serving.

Just before serving, sprinkle the sugar over the top of the brulées and then grill them under a high heat until the sugar dissolves and caramelizes to leave a glass-like topping.

opposite: preserved lemon granita, *following page:* passion fruit and preserved lemon brûlées

preserved lemon and cardamom surprise pudding

The preserved lemons here provide a delicious tangy, salty flavour. Indeed the dish gets its name from the wonderful pool of lemon syrup found in base of the dish. It is best eaten straight out of the oven, dusted with icing (confectioners') sugar.

Preparation time: 20 minutes, cooking time: 40 minutes, serves 4

2 oz (60 g) butter, softened
¾ cup (6 oz/185 g) caster (superfine) sugar
3 eggs, separated
1 teaspoon grated lemon rind
1 tablespoon finely chopped, rinsed, preserved lemon rind (see page 14)
⅓ cup (1¼ oz/40 g) self-raising flour
½ teaspoon ground cardamom
¼ cup (2fl oz/60 ml) lemon juice
¾ cup (6fl oz/185ml) milk
icing (confectioners') sugar, to dust

Preheat the oven to 350°F (180°C/Gas Mark 4). Lightly brush a 4-cup (1-litre) capacity ovenproof dish with melted butter.

Beat the butter and sugar until light and creamy, add the egg yolks and lemon rinds and beat until combined.

Sift the flour and cardamom into a bowl, then fold this into the butter mixture with the lemon juice and milk. Beat the egg whites in a clean, dry bowl until soft peaks form, then fold them into the pudding mixture.

Spoon the mixture into the prepared dish and place the dish into a roasting tin. Fill the tin with enough hot water to come halfway up the side of the prepared dish and bake for 40 minutes or until risen and golden. Dust with icing (confectioners') sugar before serving.

pepper

pepper varieties

DRIED GREEN PEPPER Green peppercorns are the unripe berries of a tropical vine, from which both black and white peppers are produced. Dried and crushed they make an aromatic addition to sauces, rubs and marinades. The mild fresh flavour teams well with tomatoes, citrus fruits and cheese.

DRIED PINK PEPPERCORNS These are not true peppercorns; they come from the shinus tree and lack the peppery aroma that true pepper contains. Instead they impart a mild pine-like flavour to dishes. Pink peppercorns work well in both sweet and savoury dishes, especially cream-based sauces.

BLACK PEPPER Black peppercorns have been picked green and dried in the sun. Probably the most widely used of all pepper, black pepper is delicious with red meat and oily fish such as salmon and mackerel. It is best purchased whole and cracked just before using.

DRIED MIXED (RAINBOW) PEPPERCORNS This is a combination of the black, white, pink and green peppercorns and gives a delicious, "complete" pepper flavour. Well suited to peppermills, it works well rubbed into meats to form a crust as it adds colour, and is nice mixed with breadcrumbs for a coating for meats.

WHITE PEPPER This is made by soaking the nearly ripe peppercorns in water, then rubbing them to remove the skin before drying them. White pepper is hotter and more pungent than black, but lacks the stronger distinctive pepper flavour. The flavour teams well with seafood and garlic.

BOTTLED (PICKLED) PEPPERCORNS These are green or pink peppercorns that have been pickled in a vinegar or brine solution. Best known for their use in pepper sauce, they have a slightly tangy peppercorn flavour. They go well in creamy sauces, pastries, stuffings and with cooked meat.

DRIED SZECHUAN PEPPER Not a true pepper, this is the berry of the prickly ash tree and it is used extensively in Asian cooking. Be sure to roast it before using to bring out the flavour. It is delicious added to stir fries, batters and combined with salt and sprinkled on fried foods.

FRESH GREEN PEPPER Strands of fresh green peppercorns are available from Asian food stores. Subtler in flavour than the dried peppercorns, they are perfect in curries, stir fries and dressing. They can be used wherever dried or bottled peppercorns are indicated in recipes. They freeze well too.

cajun pepper

Put 2 tablespoons of cayenne pepper, 2 tablespoons of paprika, 2 tablespoons of garlic powder, 1 tablespoon of dried oregano, 1 tablespoon of dried thyme, 1 tablespoon of onion powder, 1 tablespoon of fine sea salt, 1 teaspoon of white pepper, and 1 teaspoon of cracked black pepper into a bowl and mix to combine. This is great as a rub for meat or chicken that is going to be barbecued. It is also delicious sprinkled on popcorn or fries.
Makes ½ cup

parmesan, pepper and rosemary sprinkle

Put ½ cup of finely grated Parmesan, 1 teaspoon pink sea salt flakes, 2–3 teaspoons of cracked mixed peppercorns, and 1 tablespoon of chopped fresh rosemary into a bowl and mix to combine. Use this sprinkle on top of salads, steamed green vegetables, pasta and roasted vegetables or layered in potato bakes.
Makes ½ cup

citrus pepper wet rub

Put 2 tablespoons of lime juice, 1 tablespoon of honey, 2 crushed cloves of garlic, 2 teaspoons of grated lime zest, 1 teaspoon of grated lemon zest, 2 teaspoons cracked black peppercorns, and 1 tablespoon of olive oil into a bowl and mix to combine. Add fish, poultry or lamb to the mixture and allow to marinate for 30 minutes. Then simply cook in a stir fry.
Makes approx ⅓ cup

smokey garlic pepper rub

Put 2 crushed cloves of garlic, 2 tablespoons of grey sea salt (sel gris de Guerande), 2 teaspoons of crushed dried pink peppercorns, ½ teaspoon of smoked Spanish paprika, and 3 tablespoons of olive oil into a bowl and mix to combine. Rub the garlic salt mixture over both sides of whatever meat, poultry or fish you have chosen to use it on and allow it to marinate for 30 minutes before cooking.
Makes ⅓ cup

japanese seaweed and pepper sprinkle

Put ½ sheet finely chopped nori (dried seaweed), 2 tablespoons of toasted sesame seeds, ½ teaspoon of chilli flakes, 1 tablespoon of sea salt, and 2 teaspoons of cracked dried pink peppercorns into a bowl and mix to combine. This is delicious sprinkled over steamed rice, sprinkled onto deep fried foods or mixed with Japanese breadcrumbs and used as a coating for chicken or pork schnitzels.
Makes ½ cup

chicken and pink peppercorn pâté

The secret to a good pâté is in the creamy texture – it is really important not to overcook the chicken livers, otherwise it will be chalky. The apples add an extra sweetness to this recipe, which contrasts beautifully with the pink peppercorns.

Preparation time: 15 minutes + 4 hours chilling, cooking time: 10 minutes, serves 6

3 oz (100 g) butter
4 shallots, finely chopped
1 apple, peeled and grated
1 lb (500 g) chicken livers
sea salt flakes and cracked black pepper
2 tablespoons dried pink peppercorns
3 tablespoons brandy or Cognac
2 tablespoons single (light) cream
black rice (sesame) crackers, to serve

Melt 1 oz (30 g) of the butter in a frying pan, add the shallots and cook for 3 minutes or until soft. Add the apple and cook for 3 minutes or until soft. Transfer the mixture to a food processor.

Melt another 1 oz (30 g) butter in the same frying pan, add the chicken livers, then season with salt and pepper. Next, add the pink peppercorns and cook over a medium heat for 2 minutes or until browned on the outside and pink inside. Add to the food processor.

Put the brandy in the frying pan and light with a long match. When the flames have died down, pour this into the food processor along with the cream. Process until smooth.

Cut the remaining butter into cubes. With the food processor motor running, add the butter a cube at a time and process until combined. Pour into 4 ramekins, cover the surface of each with plastic wrap and refrigerate until firm. Serve the pâté with black rice (sesame) crackers.

fresh yogurt balls with chilli pepper coating

Labna – as these fresh yogurt balls are better known in the Middle East – are great served with crackers or as part of a meze or antipasto platter. For a sweet treat, roll them in chopped nuts with pepper and drizzle honey over the top.

Preparation time: 40 minutes + 3 days chilling and draining, makes 20 balls

2 clean 20 in x 20 in (50 cm x 50 cm) muslin squares

6 cups plain Greek-style yogurt
2 teaspoons sea salt flakes
2 teaspoons dried chilli flakes
2 tablespoons dried mixed (rainbow) peppercorns, cracked
1 teaspoon pink sea salt flakes

Put the yogurt and salt into a bowl and mix to combine. Put the muslin squares on top of each other and place the yogurt mixture in the center. Gather up the corners, tie firmly with string and suspend the bag over a bowl. Refrigerate for 3 days.

Once drained, remove the cloth and discard. Combine the chilli flakes, pepper and salt then shape tablespoons of the yogurt into balls and roll these in the seasoning mixture. Try serving the balls with rocket (arugula) on crisp Italian bread.

* To store: fill a 4-cup (1-litre) jar with 2 cups (16 fl oz/500 ml) olive oil and put the balls into the jar. They will last for about 1 week if refrigerated, but will need to be kept at room temperature for about an hour before using in order for the oil to soften.

peppered tuna with wasabi mayonnaise

Make sure you do not overcook the tuna as it needs to be rare for this recipe – if it is cooked for too long it will make slicing difficult. The oily tuna cuts through the pungency of the pepper well.

Preparation time: 10 minutes + 10 minutes cooling, cooking time: 2 minutes, serves 4–6 as an appetizer

2 tablespoons soy sauce
2 teaspoons lemon juice
10 oz (300 g) tuna fillet
1½ tablespoons cracked black pepper
1 lime, finely grated
1 tablespoon vegetable oil
½ teaspoon wasabi paste
⅓ cup whole egg mayonnaise

Combine the soy sauce and lemon juice then cut the tuna into 1¼-in (3-cm) thick strips and place them into the liquid mixture. Combine the pepper and lime then roll the tuna in the mixture to coat it with a lime pepper crust.

Heat the oil in a frying pan then, once it has reached a medium-high heat, add the tuna and cook for 1 minute on each side. Remove the tuna from the pan and immediately wrap it firmly in plastic wrap in order to seal the juices in. Set aside and allow to cool.

Use a sharp knife to cut the cooled tuna into thick slices. Combine the wasabi paste and mayonnaise and serve with the tuna slices.

bloody mary

This is a favourite drink of mine, probably because it feels slightly healthy and more like a meal than a beverage. Omit the vodka and serve it as a refreshing drink for brunch, the colder the better.

Preparation time: 5 minutes, serves 4

2 cups crushed ice
½ cup (4 fl oz/125 ml) vodka
4 cups (1 litre) tomato juice
2 teaspoons celery salt
2 teaspoon cracked black pepper
Tabasco sauce, to taste
Worcestershire sauce, to taste
4 stalks celery

Fill 4 tall glasses with crushed ice. Divide the vodka between the glasses. Top with the tomato juice, half the celery salt and half the pepper and stir well to combine. Season to taste with a few drops of Tabasco and a few drops of Worcestershire sauce. Stir again to combine.

Cut small 2-in (5-cm) slits in the top of each celery stick and then press these into the celery salt and pepper mixture. Stand the celery sticks in the glasses just before serving.

opposite: peppered tuna with wasabi mayonnaise, *following page:* bloody mary

pepper parmesan shortbread

These mouthwatering shortbreads are wonderful bites to serve with drinks. The recipe uses short, buttery pastry. If making in the summer, you may need to rest the pastry in the refrigerator for half an hour before rolling it out.

Preparation time: 30 minutes, cooking time: 25 minutes, makes approx 35

1¼ cups (5 oz/155 g) plain (all-purpose) flour
6 oz (180 g) butter, chilled
1 cup (3½ oz./100g) finely grated Parmesan cheese
1 tablespoon bottled (pickled) green peppercorns, chopped

Preheat the oven to 350°F (180°C/Gas Mark 4). Put the flour and butter into a bowl and rub the butter into the flour until the mixture resembles fine breadcrumbs. Stir in the Parmesan and peppercorns. Use your fingers to bring the mixture together into a ball, then turn out onto a lightly floured surface.

Roll the dough out between 2 sheets of baking paper. Cut out the shortbread shapes using shaped cutters and arrange onto 2 baking trays lined with paper. Bake for 25 minutes or until crisp and golden.

* The shortbreads can be made up to 1 day in advance and stored in an airtight container. Alternatively, the dough can be wrapped in plastic wrap and frozen.

thai sweet prawn and pepper soup

This soup wonderfully combines fiery white peppercorns with the sweetness of the tomatoes and palm sugar. The coriander adds a lovely freshness to the dish – Asian food stores should sell coriander with the roots still attached.

Preparation time: 20 minutes, cooking time: 25 minutes, serves 4

3 cloves garlic, peeled
1 tablespoon fresh coriander root, chopped
pinch sea salt
10 white peppercorns
1 tablespoon groundnut (peanut) oil
3 tablespoons grated palm sugar or brown sugar
3 tablespoons nam pla (Thai fish sauce)
3 cups (24 fl oz/750 ml) chicken or fish stock
2 tomatoes, cut into thin wedges
1 lb (500 g) raw prawns (shrimps), peeled and deveined with the tails left intact
fresh coriander (cilantro) leaves, to serve
½ cup fried shallots, to serve
ground white pepper, to serve

Put the garlic, coriander root, salt and pepper into a mortar and pestle or spice grinder and grind to form a smooth paste. Heat the oil in a wok and the paste and cook over a medium heat for 3 minutes or until golden. Add the palm (or brown) sugar and cook, stirring constantly, until the sugar melts. Add the fish sauce, stock and tomatoes, reduce the heat and simmer for 15 minutes. Add the prawns (shrimps) and simmer for 3–5 minutes or until tender.

Divide the soup between 4 serving bowls and top with the coriander (cilantro) leaves, shallot and pepper just before serving.

steamed asparagus with basil pepper mayonnaise

The mayonnaise is delicious served with vegetables as an appetizer (as shown here) or it can also be served as an accompaniment with cooked fish, prawns or chicken. Use a good-quality mayonnaise or, if you have time, you could make your own.

Preparation time: 10 minutes, cooking time: 5 minutes, serves 4 as an appetizer

10 oz (300 g) asparagus spears
1 cup fresh basil leaves
2 cloves garlic
½ teaspoon sea salt flakes
1 teaspoon black peppercorns
½ cup whole egg mayonnaise

Blanch or steam the asparagus until bright green and tender. Drain and refresh in iced water.

Put the basil, garlic, salt and pepper into a mortar and pestle and crush until smooth. Fold the purée through the mayonnaise then serve the asparagus spears with the mayonnaise.

 opposite: thai sweet prawn and pepper soup, *following page:* steamed asparagus with basil pepper mayonnaise

peppered rösti with smoked salmon

Try to make sure that you squeeze every last drop of moisture out of the potatoes, as this will give nice crispy rösti cakes. Don't have the oil and butter over too high a heat or the outside of the rösti will burn before the potato cooks.

Preparation time: 25 minutes, cooking time: 15 minutes, serves 4 (makes 12)

2 eggs, lightly beaten
3 tablespoons plain (all-purpose) flour
1 tablespoon chopped fresh chives
1½ teaspoons dried mixed (rainbow) peppercorns, crushed
2 tablespoons finely grated Parmesan cheese
1 lb (500 g) potatoes, peeled
1 teaspoon fleur de sel (French sea salt)
2 tablespoons olive oil
¾ oz (20 g) butter
3½ oz (100 g) gravlax (see page 18) or smoked salmon, to serve
crème fraîche, to serve
8 caperberries, to serve

Whisk together the eggs, flour, fresh chives, peppercorns and Parmesan cheese. Grate the potatoes and place them in a clean dry tea (dish) towel. Squeeze firmly to remove as much moisture as possible from the potatoes then add to the egg mixture and mix to combine. Season with the salt.

Heat the oil and butter in a frying pan. Add 2 tablespoons of the batter mixture at a time and cook over a medium heat, flattening each to ensure it cooks evenly and is crisp and golden brown on both sides. Remove, drain on absorbent kitchen paper and keep warm while you cook the remaining batter.

Serve the peppered rösti topped with the gravlax or smoked salmon, crème fraîche and caperberries.

avocado green salad with pink peppercorn vinaigrette

This salad makes a light, fresh start to any meal. Use whichever greens you can find – I like the bitter leaves shown here. If you are entertaining and want to dress the dish up a little more you can top it with some chargrilled prawns or scallops.

Preparation time: 15 minutes, serves 4 as an appetizer

2 large avocados, halved
¾ oz (25 g) watercress sprigs
¾ oz (25 g) lamb's lettuce or baby rocket (arugula)

For the vinaigrette:
1 clove garlic, crushed
1 teaspoon dried pink peppercorns, crushed
1 teaspoon honey
1 teaspoon Dijon mustard
1½ tablespoons red wine vinegar
¼ cup (2 fl oz/60 ml) extra virgin olive oil

Arrange the avocado halves onto individual serving plates. Put the watercress and lettuce or rocket (arugula) into a bowl and gently toss together to combine.

Fill each hole in the avocado with the lettuce mix. Whisk together the garlic, peppercorns, honey, mustard, vinegar and olive oil. Drizzle the vinaigrette over the lettuce and avocado and serve immediately.

pink pepper and shallot relish

This relish is delicious teamed with vintage Cheddar cheese and served on crackers, or you can serve it with barbecued meats.

Preparation time: 5 minutes, cooking time: 40 minutes + 1 hour cooling, makes approx 4 cups (1 litre)

2 lb (1 kg) shallots, peeled and halved
1½ cups (12 fl oz/375 ml) malt vinegar
1½ cups brown sugar
2 tablespoons dried pink peppercorns
3 black peppercorns
1 bay leaf
1 tablespoon fresh lemon thyme leaves

Put the shallots, vinegar, sugar, peppercorns, bay leaf and thyme leaves into a saucepan and simmer for 40 minutes or until the mixture thickens and there is hardly any moisture left. Spoon into sterilised jars and seal.

* The relish will keep for up to 3 months.

 opposite: avocado green salad with pink peppercorn vinaigrette, *following page:* pink pepper and shallot relish

peppery vietnamese beef noodle soup

Travelling in Vietnam a few years ago, I began a long-lasting, loving relationship with pho – Vietnamese noodle soup. If the uncooked beef is a little off-putting, just add it to the stock and cook it for a few minutes before serving.

Preparation time: 15 minutes,
cooking time: 25 minutes + 10 minutes standing,
serves 4

6 cups (1.5 litres) good-quality beef stock
3½ oz (100 g) fresh ginger, peeled, sliced and bruised
2 cinnamon sticks
2 star anise
1 teaspoon black peppercorns
3 tablespoons nuoc man (Vietnamese fish sauce)
6½ oz (200 g) dry rice noodles
3½ oz (100 g) bean sprouts
7 oz (225 g) beef fillet, very thinly sliced
coriander (cilantro) leaves, strips of spring onions (scallions), lime wedges and finely cracked black pepper, to garnish

Put the stock, ginger, cinnamon, star anise, black peppercorns and fish sauce into a large saucepan, bring to a boil then reduce the heat, cover and simmer for 20 minutes. Strain and discard the seasonings, return the stock to the pan and bring the pan back up to a boil.

Place the rice noodles in a separate saucepan and cover with boiling water. Allow to stand for 10 minutes until soft. Then put the noodles, bean sprouts and thinly sliced beef into bowls, ladle the stock over the top and serve garnished with the coriander (cilantro) leaves, spring onions (scallions), lime and pepper.

deep fried fish with garlic and pepper

If I had to choose a Thai dish closest to my heart it would be this. The coriander root is an essential part of the recipe as it too has a wonderful peppery flavour. It stores well in the freezer, which means you can have a supply always to hand.

*Preparation time: 10 minutes +
1 hour marinating,
cooking time: 7–10 minutes, serves 4*

1 lb 10 oz (800 g) snapper or Dover sole
4 cloves garlic, roughly chopped
2 tablespoons fresh coriander root, chopped
¼ teaspoon sea salt flakes
1 teaspoon white peppercorns
1 teaspoon fresh green peppercorns
groundnut (peanut) oil, for deep frying

Cut 3–4 deep slits in the thickest part of the fish on both sides. Pat it dry with absorbent kitchen paper. Put the garlic, coriander root, salt and peppercorns into a mortar and pestle or spice grinder and grind to form a smooth paste. Rub the paste over both sides of the fish, cover and allow to marinate for 1 hour.

Heat the oil in a wok then deep fry the fish for 7–10 minutes or until crisp and golden and cooked through. Drain on absorbent kitchen paper before serving. It is nice accompanied by wedges of lime.

peppered pork chops with sweet vinegar pears

The longer the chops marinate in the honey, paprika, pepper and rosemary the better. Don't worry if you can't find corella pears as beurre bosc will do the job, although you may have to double the wine, sugar and water mixture and extend the cooking time.

Preparation time: 20 minutes + 4 hours marinating, cooking time: 20–25 minutes, serves 4

4 pork chops, trimmed
2 tablespoons honey
1 teaspoon paprika
1 tablespoon dried mixed (rainbow) peppercorns, crushed
2 teaspoons fresh rosemary, chopped
2 tablespoons olive oil
8 small pears, preferably corella, halved
1 tablespoon white wine vinegar
1 tablespoon soft brown sugar

Preheat the oven to 400°F (200°C/Gas Mark 6). Put the pork chops into a shallow dish. Combine the honey and paprika then rub this over the pork. Combine the peppercorns and rosemary and then sprinkle over each side of the chops. Allow to marinate for 4 hours.

Heat the oil in a frying pan, add the pork chops and cook over a medium-high heat for 2 minutes on each side or until browned. Transfer to a baking dish.

Put the pears into a bowl, add the vinegar, sugar and 1 tablespoon of water and mix to coat the pears. Add the pears to the baking dish, cover with foil and bake for 10–15 minutes or until the pears and pork are just tender. Remove the foil and cook for a further 5 minutes, allowing the pork and pears to brown.

duck breast with red wine and pepper sauce

In this recipe I have chosen to use duck breasts that have had their skins removed – if you would like to leave the skin on be sure to prick it with a fork to give you a nice crisp skin. Use a good-quality red wine, as it is the main ingredient.

Preparation time: 15 minutes, cooking time: 30–35 minutes + 10 minutes standing, serves 4

1 tablespoon olive oil
4 duck breasts

For the sauce:
4 shallots, finely chopped
10 black peppercorns
1 sprig thyme
1 bay leaf
2 cups (16 fl oz/500 ml) red wine
1 cup (8 fl oz/250 ml) chicken stock
¼ oz (10 g) butter
1 teaspoon cracked black pepper
steamed green vegetables, to serve

Preheat the oven to 400°F (200°C/Gas Mark 6). Heat the oil in a frying pan, add the duck breasts and cook over a medium heat for 5 minutes or until crisp and browned. Turn and cook the other side for 3 minutes or until browned then transfer to a baking tray, reserving the liquid in the frying pan for the sauce. Bake the duck for 15 minutes or until tender. Remove from the oven, cover and allow to stand while you prepare the sauce.

To make the sauce, remove all but a tablespoon of the liquid from the frying pan, add the shallots and cook over a medium heat for 3 minutes or until soft. Add the peppercorns, thyme, bay leaf and red wine. Bring to a boil and cook over a high heat for 5 minutes or until the pan is almost dry. Add the stock and bring back to a boil, scraping the base of the pan to remove any sediment. Strain the sauce, return it to the heat, add the butter and pepper and cook until the butter melts. Serve it poured over the duck breasts, and accompany the dish with steamed green vegetables.

thai beef salad with peppercorn curry dressing

This dish is not as fiery as you may expect – the fresh peppercorns are not that hot and there is a lot sugar in the dressing, which helps to temper the heat of the curry paste. It is a delightful combination.

Preparation time: 20 minutes, cooking time: 12–15 minutes + 10 minutes standing, serves 4–6

1 lb (500 g) lean rump steak
3½ oz (100 g) mixed salad leaves
3 tomatoes, chopped
1 Lebanese cucumber, sliced
1 carrot, sliced
1 large red chilli, seeded and thinly sliced
½ cup fresh mint leaves

For the dressing:
2 tablespoons fresh green peppercorns, lightly crushed
1 tablespoon red curry paste
1 cup (8 fl oz/250 ml) coconut milk
2 tablespoons nam pla (Thai fish sauce)
3 tablespoons grated palm sugar or brown sugar
1 tablespoon lime juice

Cook the steak on a lightly oiled griddle (chargrill plate) over a medium-high heat for 5 minutes on each side or until medium rare. Cover and allow the steak to stand for 10 minutes before cutting it into thin slices.

Arrange the salad leaves on a large platter, top with the tomatoes, cucumber, carrot, chilli and the mint. Place the steak slices on top.

To make the dressing, put the peppercorns, curry paste, coconut milk, fish sauce, sugar and lime juice in a small saucepan and heat until the mixture boils and thickens. Then simply drizzle it over the salad before serving.

stir fried beef with mangetout and lemon pepper

Marinating the beef overnight is the key to this recipe, while roasting the peppercorns develops their flavours. When combined, the 2 peppers create a classic Asian seasoning.

Preparation time: 20 minutes + overnight marinating, cooking time: 10 minutes, serves 4

1 lb (500 g) rump steak
4 tablespoons dry sherry
2 tablespoons oyster sauce
2 tablespoons soy sauce
2 tablespoons grated palm sugar or brown sugar
1 teaspoon sesame oil
2 tablespoons white peppercorns
1 teaspoon Szechuan peppercorns
1 teaspoon sea salt flakes
3 tablespoons vegetable oil
1 tablespoon grated fresh ginger
2 cloves garlic, crushed
6 spring onions (scallions), sliced
6½ oz (200 g) mangetout (snow peas)
⅓ cup (2¾ fl oz/80 ml) lemon juice

Put the steak, sherry, oyster sauce, soy sauce, palm (or brown) sugar and sesame oil into a bowl, mix well to combine then cover. Place the bowl in the refrigerator and allow to marinate overnight.

Dry roast the peppercorns in a wok for 3 minutes or until fragrant. Transfer to a mortar and pestle or spice grinder and roughly crush. Add the salt and mix to combine.

Heat the oil in a wok, add the steak and cook until browned. Add the ginger, garlic and spring onions (scallions) and stir fry for 3 minutes or until the spring onion is soft. Add the mangetout (snow peas), half the salt and pepper mixture and lemon juice and stir fry for 2 minutes or until the mangetout are bright green.

Transfer to a serving dish and sprinkle with the remaining salt and pepper mixture. This beef recipe is best served just with rice.

opposite: thai beef salad with peppercorn curry dressing,
following page: stir fried beef with mangetout and lemon pepper

parmesan and green pepper veal schnitzel

If you do not have a meat mallet ask you butcher to pound the veal escalopes for you. Take care not to overheat the oil and butter, as you don't want the coating to get too dark before the meat inside is cooked.

Preparation time: 20 minutes + 30 minutes chilling time, cooking time: 15 minutes, serves 4

4 thin veal escalopes
plain (all-purpose) flour, for coating
1 egg, lightly beaten
2 tablespoons bottled (pickled) green peppercorns, chopped
1 cup fresh breadcrumbs
¼ cup finely grated Parmesan cheese
2 tablespoons olive oil
1¾ oz (50 g) butter
steamed vegetables, to serve

Place the veal between 2 sheets of plastic wrap and flatten it with a mallet so it is ¼ in (5 mm) thick.

Spread the flour out on a plate, combine the egg and green peppercorns in a bowl and then combine the breadcrumbs and Parmesan in a separate bowl. Toss the veal first in the flour until it is coated. Shake off any excess flour. Next dip the floured veal into the combined egg and green peppercorns, then coat with the breadcrumbs and Parmesan mixture. Place onto a baking tray lined with foil, cover and refrigerate for 30 minutes; this process will help the coating stick to the veal.

Heat the oil and butter in a frying pan, add the veal and cook over a medium-high heat until crisp and golden on both sides. Drain the veal on absorbent kitchen paper before serving it with steamed vegetables.

caramelized onion, trout and pepper tart

You can eat this tart hot or cold. It makes a tasty lunch or picnic dish and will store well in the refrigerator for about 2 days. The pastry is actually delicious used in both sweet and savoury recipes.

Preparation time: 40 minutes + 15 minutes chilling, cooking time: 1 hour 25 minutes, serves 6

For the pepper pastry:
2 cups (8 oz/250 g) plain (all-purpose) flour
3½ oz (100 g) butter
3 teaspoons finely cracked black pepper
1 egg

For the filling:
3 tablespoons olive oil
3 onions, thinly sliced
1 fennel bulb, sliced
3½ oz (100 g) baby spinach leaves
6½ oz (200 g) smoked trout, broken into large pieces
3½ oz (100 g) goat's cheese, sliced
3 eggs, lightly beaten
½ cup (4 fl oz/125 ml) single (light) cream

Preheat the oven to 400°F (200°C/Gas Mark 6). Place the flour, butter and pepper into a food processor and process until the mixture resembles fine breadcrumbs. With the motor still running, gradually add the egg along with 2–3 tablespoons of water and process until the mixture comes together. Gather into a ball. Roll out between 2 sheets of baking paper, to form a circle large enough to cover the base and side of a 9½-in (24-cm) tart tin. Fit the pastry into the tin and chill for 15 minutes or until firm.

Line the pastry with baking paper and fill with baking beans or rice. Bake for 15 minutes, remove the paper and beans and cook for a further 10 minutes, then reduce the oven temperature to 315°F (160°C/Gas Mark 2–3).

To make the filling, heat the oil, add the onions and fennel and cook over a medium heat for 20 minutes or until caramelized. Add the spinach and cook until it wilts. Spread the mixture over the pastry base. Arrange the trout and goat's cheese over the top. Whisk together the eggs and cream and pour over the other ingredients in the pastry case (shell). Bake for 40 minutes or until set.

lobster with green peppercorn mornay

Use a Chinese cleaver if you have one to cut the lobsters in half, otherwise sharp kitchen scissors will work. Select lobsters large enough to ensure each person will be satisfied with a half.

Preparation time: 20 minutes, cooking time: 10 minutes, serves 4

2 cooked lobsters
1¾ oz (50 g) butter
3 shallots, finely chopped
1 tablespoon bottled (pickled) green peppercorns, roughly chopped
2 tablespoons plain (all-purpose) flour
½ teaspoon Dijon mustard
300 ml (9½ fl oz) milk
2 egg yolks
3½ oz (100 g) grated Cheddar cheese
sea salt flakes

Cut the lobster in half and remove the meat. Discard the head matter (you may reserve it and add it to the sauce if you wish). Cut the lobster meat into small, bite-size pieces and return to the shell.

Melt the butter in a saucepan, add the shallots and green peppercorns and cook over a medium heat for 3 minutes or until soft. Add the flour and cook, stirring constantly, for 2 minutes or until the flour is golden. Add the Dijon mustard to the pan. Remove the saucepan from the heat and gradually stir in the milk. Return the pan to the heat and cook, stirring constantly, over a medium heat until the sauce boils and thickens.

Remove the pan from the heat, add the eggs and cheese and stir until combined. Season with the sea salt. Put the lobsters onto a baking tray. Spoon the sauce over the lobster and grill under a medium-high heat until golden brown. Serve immediately.

lamb fillets with pistachio and pepper

If you do not have a food processor you can chop the pistachios and use a mortar and pestle to roughly crush the peppercorns. You are looking to achieve a roughly chopped colourful crust for the lamb.

Preparation time: 20 minutes, cooking time: 15–20 minutes + 10 minutes standing, serves 4

¾ oz (25 g) shelled pistachios kernels
½ oz (15 g) white peppercorns
⅛ oz (5 g) dried pink peppercorns
1¾ oz (50 g) fresh breadcrumbs
4 lamb loins approx 5 oz (150 g) each
1 tablespoon wholegrain mustard
2 cloves garlic, crushed
baby spinach and baby carrots, to serve

Preheat the oven to 425°F (220°C/Gas Mark 7). Put the pistachios and peppercorns into a food processor and process until roughly chopped. Transfer to a shallow dish and add the breadcrumbs, stirring to combine.

Combine the mustard and garlic then coat the lamb in this mixture. Next roll the meat in the pepper mixture. Roast the coated lamb on a rack in a baking dish for 15–20 minutes or until cooked to your liking. Remove from the oven, cover and allow to stand for 10 minutes before slicing. Serve with baby spinach and baby carrots.

peppered salmon with herbed lemon butter

This is a simple recipe that makes a perfect evening meal. If you are running short on time instead of freezing the butter, you can soften it in the microwave (be careful not to melt it) and stir through the lemon, herbs and Parmesan.

Preparation time: 20 minutes + 30 minutes freezing time, cooking time: 10 minutes, serves 4

4 oz (125 g) butter, softened
1 clove garlic, peeled
1 teaspoon fleur de sel (French sea salt)
1 teaspoon grated lemon zest
2 tablespoons fresh herbs (dill, chives and chervil), chopped
2 tablespoons finely grated Parmesan cheese
4 salmon fillets
2 tablespoons black peppercorns, crushed
creamy mashed potato, to serve
watercress, to serve

Put the butter, garlic, fleur de sel, lemon zest, herbs and Parmesan into a food processor and process until smooth and combined. Shape into a roll, wrap in plastic wrap then place in the freezer until firm. Cut into thick slices.

Put the black peppercorns on a plate and press the salmon into them until coated. Cook under a preheated grill on high for 5–7 minutes or until cooked to your liking.

Serve the salmon with a slice of herbed butter on top and accompanied by creamy mashed potato and watercress.

sesame pepper chicken drumsticks

This recipe can be made using either chicken wings or drumsticks. Turning the drumsticks a couple of times during cooking will give you an even colour and also stop them sticking. This dish is great served hot or cold.

Preparation time: 15 minutes + minimum 4 hours marinating, cooking time: 40 minutes, serves 4–6

2 lb (1 kg) chicken drumsticks
½ teaspoon sesame oil
1 teaspoon black peppercorns, crushed
4 cloves garlic, chopped
¼ cup (2 fl oz/60 ml) dry sherry
½ cup (4 fl oz/125 ml) honey
⅓ cup (2¾ fl oz/80 ml) soy sauce
1 tablespoon nuoc man (Vietnamese fish sauce)
2 tablespoons sesame seeds

Make deep incisions in the top of each drumstick on both sides. This helps them cook all the way through. Put the sesame oil, pepper, garlic, sherry, honey, soy sauce, fish sauce and sesame seeds into a bowl and mix to combine. Add the chicken drumsticks and mix with your hands to coat the drumsticks in the sauce. Cover and refrigerate for 4 hours or overnight.

Preheat the oven to 400°F (200°C/Gas Mark 6). Transfer the drumsticks and marinade to a large ovenproof dish. Bake the drumsticks for 40 minutes, turning a couple of times during cooking, until the chicken is tender and cooked through.

pepper steak

Rib eye or sirloin steak are also suitable for this recipe; if you like your steak well done choose a cut that has a slight marbling of fat such as sirloin. The brandy and cream work together well here to mellow the heat of the 2 peppers.

Preparation time: 15 minutes + 15 minutes standing, cooking time: 25 minutes, serves 4

4 scotch fillet or sirloin steaks
2 cloves garlic, crushed
2 tablespoons olive oil
1 tablespoon whole black peppercorns

For the sauce:
¼ cup (2 fl oz/60 ml) brandy
½ cup (4fl oz/125ml) white wine
½ cup (4fl oz/125ml) beef stock
1 cup (8 fl oz/250 ml) single (light) cream
2 tablespoons bottled (pickled) green peppercorns, lightly crushed

steamed new potatoes, to serve
green salad, to serve

Trim the steak of any excess fat and sinew. Combine the garlic with 1 tablespoon of the oil and brush over both sides of the steaks.

Crush the black peppercorns in a mortar and pestle until roughly crushed. Place onto a flat plate and press the steaks into the pepper until both sides are coated. Cover and allow to stand for 15 minutes in order for the meat to absorb the flavours.

Heat the remaining oil in a frying pan until smoking, then add the steak and cook over a high heat, turning once, until cooked to your liking. (I recommend 1 minute each side for rare, 2–3 minutes extra each side for medium rare and 5–6 minutes each side for well done.) Remove from the pan, cover loosely with foil and allow to stand while you make the sauce.

Deglaze the pan with the brandy and wine and cook over a high heat, stirring to release any juices that may be stuck to the bottom of the pan. Stir in the stock, cream and peppercorns, and cook until the sauce is thick enough to coat the back of a spoon. Return the steaks and any juices to the pan briefly to reheat. Serve with steamed new potatoes and green salad.

opposite: sesame pepper chicken drumsticks, *following page:* pepper steak

ravioli with peppered pumpkin and goat's cheese

Using wonton wrappers to make ravioli is a great time saver. You can buy them in Asian food stores – they are about 4 in (10 cm) square. Don't confuse them with spring roll wrappers, which are larger.

Preparation time: 30 minutes, cooking time: 30 minutes, serves 4

1 lb (500 g) pumpkin or butternut squash, peeled and chopped
2 teaspoons dried mixed (rainbow) peppercorns, crushed
1 tablespoon fresh lemon thyme, chopped
3 spring onions (scallions), chopped
3½ oz (100 g) goat's cheese, crumbled
8 oz (250 g) packet wonton wrappers
1¾ oz (50 g) butter
½ cup (4 fl oz/125 ml) dry white wine
½ cup (4 fl oz/125 ml) chicken stock
1¾ oz (50 g) Parmesan shavings, to garnish
fresh basil leaves, to garnish
cracked black pepper, to garnish

Cook the pumpkin in a large saucepan of boiling water until soft, then rinse and drain well. Mash the pumpkin until it is smooth. Add the peppercorns, thyme, spring onions (scallions) and goat's cheese to the pumpkin and mix to combine.

Place 1 wonton wrapper onto a clean, dry work surface, and put 1 tablespoon of the pumpkin mixture into the center. Brush the edges lightly with water and top with another wonton wrapper. Press the edges with a fork to seal. Repeat the process with the remaining mixture and wrappers.

Cook the ravioli in batches in a large saucepan of boiling water for 5 minutes or until they float to the surface. Remove and transfer to a warmed serving dish.

Melt the butter in a frying pan, add the white wine and stock and bring to a boil, then cook over a high heat for 10 minutes or until reduced slightly.

Pour the sauce over the ravioli, sprinkle with the Parmesan shavings and basil and a little cracked black pepper.

cheese and pepper chicken with a zesty tomato salsa

You can make this recipe even easier for yourself by using a peppercorn-flavoured vintage Cheddar cheese. The reason I combine them myself in this recipe is because I like to add a few more peppercorns than store-bought varieties have.

Preparation time: 25 minutes, cooking time: 20 minutes, serves 4

4 chicken breast fillets
3 cloves garlic, crushed
1 cup grated vintage Cheddar cheese
2 tablespoons fresh or dried green peppercorns, crushed

For the salsa:
2 ripe tomatoes, chopped
½ small red onion, chopped
1 tablespoon jalapeño peppers, chopped
1 tablespoon lime juice
1 tablespoon extra virgin olive oil
1 tablespoon fresh coriander (cilantro), chopped

Preheat the oven to 400°F (200°C/Gas Mark 6). Cut a small pocket in the side of each chicken breast, making sure you do not cut all the way through to the other side. Combine the garlic, cheese and peppercorns and divide the filling between the pockets, securing them with cocktail sticks (toothpicks). Put on a baking tray and bake for 15–20 minutes or until tender.

To make the salsa, combine the tomatoes, onion, jalapeños, lime juice, olive oil and coriander (cilantro) in a bowl and mix to combine. Serve the chicken breast with the salsa. Mixed salad leaves make a good accompaniment.

potato and peppered salami salad

German salami comes coated with pepper and other spices. I love it in sandwiches and find it also goes particularly well in potato salads. Ask the delicatessen to cut it a little thicker than normal as chunky pieces are best for this recipe.

Preparation time: 20 minutes, cooking time: 15 minutes + 30 minutes cooling, serves 6

3 oz (100 g) peppered salami
1 tablespoon olive oil
2 lb (1 kg) baby new potatoes
6 spring onions (scallions), sliced
3 polski pickles (large gherkins), chopped
1 tablespoon fresh mint, chopped
½ cup sour cream
1 tablespoon horseradish cream
1 tablespoon wholegrain mustard
2 teaspoons honey
4 hard-boiled eggs

Cut the salami into thin strips. Heat the oil in a saucepan and cook the salami over a medium heat for 5 minutes or until crisp. Drain on absorbent kitchen paper.

Cut the potatoes in half or quarters depending on their size, then cook in a large saucepan of boiling water until just soft. Do not overcook or they will break up in the salad. Allow them to cool.

Put the potatoes, half the spring onions (scallions), pickles and mint into a bowl and mix to combine. Whisk together the sour cream, horseradish, mustard and honey. Add these to the potatoes and mix to combine. Grate the hard-boiled eggs over the top, then sprinkle with the salami and remaining spring onions.

watermelon, rosewater and pink peppercorn granita

This granita makes a refreshing end to a meal or can be used as a palate cleanser between courses. Alternatively, mix it with a splash of vodka or gin and serve it as a drink. The subtle pink peppercorns team perfectly with the watermelon.

Preparation time: 10 minutes + 8 hours freezing, serves 4

1¼ lb (600 g) watermelon
1 teaspoon rosewater
2 teaspoons dried pink peppercorns, roughly crushed
1 tablespoon lime juice
zest of 1 lime

Put the watermelon, rosewater, peppercorns, lime juice and lime zest into a food processor and process until combined. Pour the mixture into a shallow non-stick metal container, cover and freeze for 2 hours or until the edges are firm. Remove from the freezer, scrape the mixture with a fork to break up the crystals then return to the freezer for 1 hour. Repeat the scraping and freezing 5 times over the next 5 hours or until the ice crystals are even in size.

Scrape the granita with a fork to break it up before serving.

lime pepper coconut syrup cake

I have used the combination of lime and pepper in savoury dishes but had never before thought of using it in desserts and cakes. This cake will hopefully convert you as it did me. You can make 1 large cake or individual ones as shown here.

Preparation time: 15–20 minutes, cooking time: 55 minutes for the large cake, 40 minutes for the muffins, cake serves 8, individual muffins serve 6

4 oz (125 g) butter
2 teaspoons grated lime rind
2 teaspoons cracked black pepper
1 cup (8 oz/250 g) caster (superfine) sugar
4 eggs
2 cups (6 oz/180 g) desiccated coconut
1 cup (4 oz/125 g) self-raising flour

For the syrup:
1 cup granulated sugar
2 tablespoons lime juice
zest of 1 lime

Preheat the oven to 315°F (160°C/Gas Mark 2–3). Grease and line an 8-in (20-cm) square cake tin or six 1-cup (8 oz-/250 g-) capacity muffin holes.

Put the butter, lime rind, pepper and sugar into a bowl and beat until light and creamy.

Gradually add in the eggs, beating well after each addition. Next add the coconut and flour, folding them in until combined. Spoon the mixture into the prepared tin and bake for 50 minutes for the large cake and 35 minutes for the muffins – or until a skewer comes out clean when inserted into the center.

To make the lime syrup, place the sugar, lime juice, 1 cup (8 fl oz/250 ml) of water and lime zest into a saucepan, then stir over a low heat until the sugar dissolves. Bring to a boil and cook for 5 minutes or until the syrup has reduced and thickened.

Pour the hot syrup over the hot cake or cakes and serve immediately. This dessert marries particularly well with after-dinner coffee such as espresso.

vanilla pepper strawberries with mascarpone

Select sweet small strawberries in season for this recipe. You will find that the pepper enhances the sweetness of the strawberries to create a delicious desert. If you want something a little more colourful, use a selection of mixed berries.

Preparation time: 15 minutes,
cooking time: 5 minutes + 30 minutes cooling time + 30 minutes macerating

1 lb (500 g) small strawberries
1 vanilla pod (bean), split
1 tablespoon dried mixed (rainbow) peppercorns, cracked
½ cup caster (superfine) sugar
½ cup (4 fl oz/125 ml) Drambuie or brandy
8 oz (250 g) mascarpone cheese

Put the strawberries into a bowl and set aside. Put the vanilla pod (bean), mixed peppercorns, sugar and Drambuie or brandy into a saucepan and stir over a low heat until the sugar dissolves. Bring to a boil and cook over a high heat for 5 minutes or until the syrup reduces and thickens slightly. Allow to cool.

Pour the cooled mixture over the strawberries and set aside and allow to macerate for 30 minutes. Serve individual bowls of the strawberries with the mascarpone.

peppered chai

After visiting India I am absolutely hooked on chai. I prefer not to include black tea in my chai but if you feel as though you want a little caffeine lift add a teaspoon or two in with your milk. The pepper adds extra zest to this recipe.

Preparation time: 5 minutes, cooking time: 5 minutes + 5 minutes infusion, makes 2 cups

½ oz (15 g) whole black peppercorns
½ oz (15 g) cinnamon sticks
⅛ oz (5 g) cloves
½ oz (15 g) green cardamom pods
2 cups (16 fl oz/500 ml) milk
1 tablespoon soft brown sugar

Put the black peppercorns, cinnamon sticks, cloves and cardamom pods into a spice grinder. Grind to a fine powder.

Add 1 teaspoon of the ground spice powder to the milk, place in a saucepan and heat until it is just about to boil. Remove the pan from the heat and allow the flavours to infuse for 5 minutes. Return to the heat, add the sugar and stir until it dissolves. Strain then serve the chai immediately.

poached peaches with maple pepper cream

This simple summer recipe utilizes a glut of peaches. Be sure to use pure maple syrup as its rich, syrupy flavour marries beautifully with the fresh mint and pepper tones.

Preparation time: 15 minutes + 10 minutes standing, serves 4

4 ripe peaches
1 cup (8 fl oz/250 ml) whipping cream
1 tablespoon mint, finely shredded
1 teaspoon dried pink peppercorns, crushed
2 tablespoons maple syrup

Cut a small cross in the top of each peach. Put the peaches into a bowl, pour boiling water over them and allow to stand for 10 minutes or until the skins start to peel away. Remove the skins completely.

Beat the cream until soft peaks form. Add the mint, peppercorns and maple syrup and fold through until combined. Serve the peaches with the cream on the side.

opposite: peppered chai, *following page:* poached peaches with maple pepper cream

vanilla, pistachio and pepper ice creams

This recipe can also be made into 1 large tub of ice cream – simply fold the chopped chocolate through with the other ingredients. The flavours here work well together – the richness of the cream mellows the pepper.

Preparation time: 20 minutes, cooking time: 15 minutes + 15 minutes infusing + overnight freezing + 5 minutes standing, serves 6

1 vanilla pod (bean)
1½ teaspoons black peppercorns, crushed
½ cup (4 fl oz/125 ml) milk
½ cup pistachio kernels
13 fl oz (400 ml) can sweetened condensed milk
3 cups (24 fl oz/750 ml) double (heavy) cream

Split the vanilla pod (bean) down the center and scrape the seeds out into a saucepan. Add the pod, peppercorns and milk and heat until the mixture is about to come to a boil. Remove and allow to infuse for 15 minutes. Take the vanilla pod out of the mixture.

Put the pistachios in a food processor and process until well chopped. Add the condensed milk, cream and pistachios to the infused milk and mix to combine.

Divide the mixture between 6 x ¾ cup- (6 fl oz-/185 ml-) capacity ramekins and freeze until firm. Remove and allow to stand at room temperature for 5 minutes before serving.

aniseed, pepper and nut tart

This sticky baklava style tart is totally decadent. The subtle aniseed flavour comes from the ouzo, while the pepper cuts through the sweetness of the honey and dates. It is a rich dessert – without being sickly sweet.

Preparation time: 30 minutes, cooking time: 45 minutes + 1 hour cooling, serves 8–10

3½ oz (100 g) fresh dates, pitted and halved
1 fl oz (30 ml) ouzo
8 sheets filo pastry
1¾ oz (50 g) butter, melted
5 fl oz (155 ml) double (heavy) cream
2 oz (60 g) butter
¼ cup honey
½ cup (4 oz/115 g) soft brown sugar
1½ teaspoon black peppercorns, crushed
2 cups (6 oz/180g) flaked almonds
3½ oz (100 g) chopped walnuts
1¾ oz (50 g) pine kernels
single (light) cream, to serve

Preheat the oven to 350°F (180°C/Gas Mark 4). Soak the dates in the ouzo while you prepare the tart. Grease a 10-in (25-cm) loose-bottomed tart tin.

Brush 1 sheet of pastry with melted butter and fit it into the base of the tin, then continue layering the pastry to fit the tin. Trim the edges and discard.

Put the cream, butter, honey, brown sugar and pepper into a saucepan, bring to a boil, reduce the heat and simmer for 10 minutes. Add all the nuts and mix to combine.

Arrange the dates over the pastry base, top with the nut mixture and smooth the surface. Bake for 35 minutes or until golden brown. Cool completely then serve with cream, as shown here, or ice cream or custard.

* The fresh dates can be substituted with dried dates or figs.

pepper, rose and almond praline

I like to serve this praline to the side, either with a cuppa or to dress up a dessert. The pink peppercorns add great colour and also cut through the sweetness. Store the praline in an airtight container for up to 3 days.

Preparation time: 5 minutes,
cooking time: 20 minutes, serves 4 as a snack

4 oz (125 g) raw almonds
1 teaspoon dried pink peppercorns, crushed
1 teaspoon poppy seeds
1 cup granulated sugar
1 tablespoon fresh rose petals

Preheat the oven to 350°F (180°C/Gas Mark 4). Put the almonds onto a baking tray and roast for 10 minutes or until golden.

Line a baking tray with a sheet of baking paper. Sprinkle the almonds, peppercorns and poppy seeds over the paper.

Put the sugar and 3 tablespoons of water into a heavy-based saucepan. Cook over a low heat until the sugar dissolves then boil for 5–10 minutes or until the sugar turns a deep caramel colour. Quickly pour the toffee over the nuts and seeds and gently press the petals into the top of the toffee, taking care as the toffee will be very hot. Allow to set then break into large shards.

* To clean the sticky toffee residue from the bottom of the saucepan, fill it with water and return it to the heat. This will eventually dissolve the toffee.

peppered rhubarb with ginger meringues

Ensure the eggs are at room temperature, as this will allow you to beat more air into them. Use a stainless steel, copper or glass bowl to beat the egg whites in. And remember – it is nearly impossible to make meringues on a humid day.

Preparation time: 15 minutes,
cooking time: 40 minutes + 1 hour cooling,
serves 6

2 egg whites
½ cup (4 oz/125 g) caster (superfine) sugar
1 tablespoon glacé (candied) ginger, chopped
1 lb (500 g) rhubarb, cut into 2-in (5-cm) pieces
1 teaspoon black peppercorns, crushed
½ cup (4 oz/125 g) caster (superfine) sugar
1 teaspoon grated lemon zest
4 fl oz (125 ml) clotted cream, to serve

Preheat the oven to 300°F (150°C/Gas Mark 2). Line a baking tray with baking paper. Beat the egg whites in a clean, dry bowl until soft peaks form. Gradually add in the sugar, beating well after each addition until the mixture is stiff and glossy. Add the ginger and fold through.

Divide the mixture into 6 and drop large spoonfuls of it onto the baking tray. Bake for 30 minutes or until crisp and lightly golden. Turn off the oven and allow the meringues to cool down inside it.

When the meringues are cool, put the rhubarb, pepper, sugar, lemon and ½ cup (4 fl oz/125 ml) of water into a saucepan and cook over a low heat, stirring constantly until the sugar dissolves. Simmer, covered, for 10 minutes or until the rhubarb is just soft.

Serve the meringues with a dollop of clotted cream and topped with the rhubarb mixture.

 opposite: pepper, rose and almond praline, *following page:* peppered rhubarb with ginger meringues

indian spiced rice pudding

I find rice pudding to be the ultimate comfort food – this particular recipe is another Indian-inspired recipe. The Indians use pepper to aid digestion so I like to have this after a rich curry as a soothing medicine.

Preparation time: 10 minutes, cooking time: 1 hour 5 minutes, serves 4

4 cups (1 litre) milk
1 cinnamon stick
4 cloves
1 teaspoon dried pink peppercorns, crushed
3 cardamom pods, bruised
1 vanilla pod (bean), split
⅓ cup light muscovado (raw) sugar
½ cup short grain rice
2 tablespoons pistachio kernels, chopped, to serve
1 teaspoon dried mixed (rainbow) peppercorns, crushed, to serve

Place the milk in a saucepan and bring it almost to a boil, then add the spices and sugar and stir until the sugar dissolves. Add the rice and then stir for 1 minute or until the mixture returns to a boil. Reduce the heat and simmer over a low heat, stirring occasionally to prevent it from sticking to the base of the pan. Cook for 1 hour or until the rice is tender. Remove the cinnamon and vanilla. Combine the pistachio and mixed peppercorns and sprinkle them over the top of the rice pudding just before serving.

list of recipes

sprinkles and rubs

asian salt sprinkle 10
brining solution for meat, fish and chicken 10
cajun pepper 10
citrus pepper wet rub 10
gomashio (sesame salt sprinkle) 10
indian spice rub 10
japanese seaweed and pepper sprinkle 84
mediterranean rub 10
parmesan, pepper and rosemary sprinkle 10
pink pepper and shallot relish 102
preserved lemons 12
smokey garlic pepper rub 84
szechuan sprinkle 10
tapenade 34

drinks

bloody mary 90
margarita 42
peppered chai 144
vodka pomegranate crush with salted stirrers 30

starters and hors d'oeuvres

avocado green salad with pink peppercorn vinaigrette 102
boiled tea eggs with 5 spice sesame salt 23
chargrilled prawns with a salty thai dipping sauce 40
chicken and pink peppercorn pâté 86
fresh yogurt balls with chilli pepper coating 88
gravlax 16
ham, pecorino pepato and egg pan fried sandwich 39
marinated feta 18
olive and rosemary bread 34
olive, tomato and bacon corn bread 46
pan fried halloumi with lemon 14
pepper parmesan shortbread 90
peppered rösti with smoked salmon 100
peppered tuna with wasabi mayonnaise 90
pink salt pretzels 28
pissaladiere 60
potato and peppered salami salad 136
roast vegetable salad with preserved lemon vinaigrette 64
saffron salt cod fritters 30
salt and pepper squid 26
salted fish dip 42
salted spice and hazelnut dip 23
spanokopita 58
steamed asparagus with basil pepper mayonnaise 96

soups

miso soup 34
peppery Vietnamese beef noodle soup 106
thai sweet prawn and pepper soup 96

pasta

ravioli with peppered pumpkin and goat's cheese 132
spaghetti with chorizo, anchovies, chilli and garlic 48

meat

chinese beef with salted black bean sauce 52
lamb fillets with pistachio and pepper 124
parmesan and green pepper veal schnitzel 118
pepper steak 128
peppered pork chops with sweet vinegar pears 111
roast beef fillet in salt crust 56
salty Vietnamese clay pot pork 50
stir fried beef with mangetout and lemon pepper 114
thai beef salad with peppercorn curry dressing 114
veal saltimbocca 70

poultry

cheese and pepper chicken with a zesty tomato salsa 134
chicken with sweet chilli salt crust 52
duck breast with red wine and pepper sauce 112
feta chicken with honey caper sauce 70
sesame pepper chicken drumsticks 128

fish

caramelized onion, trout and pepper tart 120
deep fried fish with garlic and pepper 108
fish baked in salt crust with caperberry sauce 64
lobster with green peppercorn mornay 122
peppered salmon with herbed lemon butter 126
potato, fennel and anchovy bake 68

desserts

aniseed, pepper and nut tart 150
indian spiced rice pudding 156
lime pepper coconut syrup cake 140
passion fruit and preserved lemon brûlées 74
pepper, rose and almond praline 152
peppered rhubarb with ginger meringues 152
poached peaches with maple pepper cream 144
preserved lemon and cardamom surprise pudding 78
preserved lemon granita 74
vanilla pepper strawberries with mascarpone 142
vanilla, pistachio and pepper ice creams 148
watermelon rosewater and pink peppercorn granita 138

index of recipes

A
aniseed, pepper and nut tart 150
asian salt sprinkle 10
asparagus, steamed with basil pepper mayonnaise 96
avocado green salad with pink peppercorn vinaigrette 102

B
beef, stir fried with mangetout and lemon pepper 114
bloody mary 90
brining solution for meat, fish and chicken 10

C
cajun pepper 10
caramelized onion, trout and pepper tart 120
chicken and pink peppercorn pâté 86
chicken, cheese and pepper with a zesty tomato salsa 134
chicken with sweet chilli salt crust 52
chinese beef with salted black bean sauce 52
citrus pepper wet rub 10

D
duck breast with red wine and pepper sauce 112

E
eggs, boiled tea with 5 spice sesame salt 23

F
feta chicken with honey caper sauce 70
fish baked in salt crust with caperberry sauce 64
fish, deep fried with garlic and pepper 108
fresh yogurt balls with chilli pepper coating 88

G
gomashio (sesame salt sprinkle) 10
gravlax 16

H
ham, pecorino pepato and egg pan fried sandwich 39

I
indian spice rub 10
indian spiced rice pudding 156

J
japanese seaweed and pepper sprinkle 84

L
lamb fillets with pistachio and pepper 124
lime pepper coconut syrup cake 140
lobster with green peppercorn mornay 122

M
margarita 42
marinated feta 18
mediterranean rub 10
miso soup 34

O
olive and rosemary bread 34
olive, tomato and bacon corn bread 46

P
pan fried halloumi with lemon 14
parmesan and green pepper veal schnitzel 118
parmesan, pepper and rosemary sprinkle 10
passion fruit and preserved lemon brûlées 74
peaches, poached with maple pepper cream 144
pepper parmesan shortbread 95
pepper, rose and almond praline 152
pepper steak 128
peppered chai 144
peppered rösti with smoked salmon 100
pink pepper and shallot relish 102
pink salt pretzels 28
pissaladiere 60
pork chops, peppered with sweet vinegar pears 111
potato and peppered salami salad 136
potato, fennel and anchovy bake 68
prawns, chargrilled with a salty thai dipping sauce 40
preserved lemon and cardamom surprise pudding 78
preserved lemon granita 74
preserved lemons 12

R
ravioli with peppered pumpkin and goat's cheese 132
rhubarb, peppered with ginger meringues 152
roast beef fillet in salt crust 56
roast vegetable salad with preserved lemon vinaigrette 64

S
saffron salt cod fritters 30
salmon, peppered with herbed lemon butter 126
salt and pepper squid 26
salted fish dip 42
salted spice and hazelnut dip 23
sesame pepper chicken drumsticks 128
smokey garlic pepper rub 84
spaghetti with chorizo, anchovies, chilli and garlic 48
spanokopita 58
steamed asparagus with basil pepper mayonnaise 96
szechuan sprinkle

T
tapenade 34
thai beef salad with peppercorn curry dressing 114
thai sweet prawn and pepper soup 96
tuna, peppered with wasabi mayonnaise 90

V
vanilla pepper strawberries with mascarpone 142
vanilla, pistachio and pepper ice creams 148
veal saltimbocca 70
vietnamese beef noodle soup, peppery 106
vietnamese clay pot pork, salty 50
vodka pomegranate crush with salted stirrers 30

W
watermelon, rosewater and pink peppercorn granita 138

The publisher would like to thank Bison Homewares, Ceramica Blue and Sunbeam Appliances for the props and appliances supplied.

Author acknowledgements Firstly I would like to thank the amazingly talented and visionary Catie Ziller for coming up with such an original idea for a book and secondly for asking me to write it. I have loved every minute of it and now have a newfound respect for salt and pepper. Never again shall I sprinkle them on willy nilly.

To the team of wonderful women I was blessed to work with on this book – a huge thank you for putting a little more into each day than was asked; your efficiency, support and passion make my job such a pleasure. So, thanks to:

Michelle Lucia darling, who let me invade her kitchen to test, then helped make sure we got the best out of each recipe; Penel Grieve my 'solid as a rock never misses a trick tester' who helped me finish the testing off in my brand spanking kitchen in my new home and the new one on the block Sarah Tildesley, my sunshine ' hello moto' home economist – what an absolute gift you were to me in London, zipping here and there to get anything needed with a smile that could light up the sky.

Thanks also to: Deirdre Rooney my divinely dedicated photographer, who point blank refuses to shoot one version of anything – thank you for being so focused on producing the best book possible. What a marvellous job you have done.

Jane Campsie, my superb props stylist and dear gifted friend, for sourcing such lovely pieces for me to work with and providing me with such a comfortable nurturing home for the duration of my stay in London.

Claire Musters, my totally onto it editor for making sure all was on time and as it should be. Your unwavering attention to detail is awe-inspiring.

Special mention to my friends Barry, Anton and Christabel for taking such fabulous care of my darling girl Pride Joy while I was in London shooting this book.

Finally to my family and friends for the big love, fat fun, unwavering support and gold-medal friendship they continue to provide me with – I am grateful each and everyday to have such special shining beings in my life.

Published in the United States and Canada in 2006 by Whitecap Books Ltd.

For more information, contact Whitecap Books Ltd., 351 Lynn Avenue, North Vancouver, British Columbia, Canada V7J 2C4. Visit our website at www.whitecap.ca.

Author and stylist Jody Vassallo
Recipe testing Michelle Lucia, Penel Grieve
Props stylists Jane Campsie
Food for photography Sarah Tildesley
Photographer Deirdre Rooney

ISBN 1-55285-816-2
ISBN 978-1-55285-816-5
Printed in Singapore